John Chiene

Lectures on the Elements

Or the First Principles of Surgery

John Chiene

Lectures on the Elements
Or the First Principles of Surgery

ISBN/EAN: 9783337277154

Printed in Europe, USA, Canada, Australia, Japan

Cover: Foto ©berggeist007 / pixelio.de

More available books at **www.hansebooks.com**

LECTURES

ON THE

ELEMENTS OR FIRST PRINCIPLES

OF

SURGERY

BY

JOHN CHIENE M.D. Edin.

F.R.S.E. F.R.C.S.E.

PROFESSOR OF SURGERY IN THE UNIVERSITY OF EDINBURGH

Reprinted from ' The American Practitioner'

EDINBURGH: DAVID DOUGLAS

1882

.

THE AMERICAN PRACTITIONER.

MARCH, 1879.

Certainly it is excellent discipline for an author to feel that he must say all that he has
to say in the fewest possible words, or his reader is sure to skip them; and in the plainest
possible words, or his reader will certainly misunderstand them. Generally, also, a downright fact may be told in a plain way ; and we want downright facts at present more than
anything else.—Ruskin.

Original Communications.

THE ELEMENTS OF SURGERY.*

BY JOHN CHIENE, M.D., F.R.C.S.E.

Surgeon to the Edinburgh Royal Infirmary, etc., etc.

Lecture I.—The Study of Principles—Health and Disease—Action of an
Irritant—The Physiology of an Injured Part—The Blood—Blood Vessels
and Tissues—The Formation of Inflammatory Lymph—The Process of Repair
—Hemorrhage—The Repair of a Wounded Vessel, and Repair in the Tissues
Generally, by Blood Clot.

Introduction.—Among the many advances made in surgical
practice during the present century, no two agents have come
to play so extensive a part as anæsthetics and antiseptics.
By the first pain is abolished ; by the second putrefaction is
prevented. Surgeons are of one mind as to the benefits and
advantages of anesthesia, but they are not agreed as to which
anesthetic is in all respects the best. There is far greater
diversity of opinion, however, as to the value and capabilities
and uses of antiseptics. Many surgeons seem unwilling to
believe that putrefaction is preventible ; others assert that the

* The several lectures in this series are an abstract of certain lectures delivered by me on Systematic Surgery in 1878, and asked for by Dr Yandell.
Much that I say is already known, and has been better said before. Many
points which it was proper to dwell upon in the class-room are omitted here
If certain portions which remain be thought too elementary for this place, I

methods in use for this end are not effective, others again that
the means are not adapted to every day work, while yet another class regards the dangers arising from putrefaction as of
too little moment to demand such an amount of care and
trouble to avert them. Having been for some years associated with Mr Lister in this hospital, and privileged to observe
his practice and not infrequently to be intrusted with his wards
when he was absent from Edinburgh, I have had abundant
opportunity of studying and putting to practical test the doctrines of which he is the exponent. The conclusions which
have forced themselves upon my mind are that putrefaction
in wounds is, as Mr Lister asserts, the result of deposit in
them of living organisms, the germs of which are present in
the air and water ; that in consequence of the presence of
these organisms certain products are formed, which act locally
as irritants and prevent healing of the wound, and which, if
absorbed, give rise to constitutional and local symptoms ; that
these in turn will vary in intensity in different individuals, the
differences depending upon, first, the variety of the putrefactive poison, second, its amount, third the constitution of the
patient. As in agriculture, so in surgery, the yield per acre

trust the excuse for their appearance will be found in the fact that they are,
in my opinion, necessary for the proper elucidation of the subject.

The authorities to which frequent reference will be made are,—first, John
Hunter's Works, edited by Palmer. Second—Anatomical and Pathological
Observations, by John and Harry Goodsir, Edinburgh, 1845. Third—Paget's
Pathology, edited by Turner. Fourth—Virchow's Cellular Pathology. Fifth—
Foster's Physiology. Sixth—Billroth's Surgery. Seventh—Burdon Sanderson
on Inflammation in Holmes's System of Surgery. Eighth—Druitt's article
on Inflammation, in Cooper's Surgical Dictionary. Ninth—On the Coagula-
tion of the Blood (Lister), Proc. Royal Society, 1863. Tenth—Contributions
to Physiology and Pathology (Lister), Philosophical Transactions for 1858.
Eleventh—Chirurgie Antiseptique, by Lucas-Championnire, Baillière, 1876.
In this work references will be found to Lister's papers on Antiseptic Surgery
in the Lancet, 1867, 1869 ; and in the British Medical Journal, October, 1868.
Twelfth—Contributions to the Germ Theory of Putrefaction by Lister, in the
Transactions of the Royal Society of Edinburgh, Vol. XXVII., and in Micro-
scopical Journal, October, 1873.

My thanks are especially due to Mr A. M. Stalker, M.A., for the care and
intelligence with which he has transcribed these lectures from notes taken while
they were being delivered.

is governed by the variety and amount of seed sown, and the soil in which it is sown. The organisms are the seed, and vary as the different varieties of the cereals vary among themselves. The tissues are the soil, and vary in their vitality—in their " power of resistance," as John Hunter termed it—as sand, loam, gravel, and clay vary.

Lister, to my mind, has clearly shown that an active faith in the germ theory of putrefaction, as taught by Pasteur, will enable surgeons to work with far better chances of success than is otherwise possible; and that by the adoption of certain methods and use of certain substances termed antiseptics, putrefaction can be absolutely prevented. If these doctrines be, as I believe, founded in truth, their final acceptance is but a matter of time. The methods by which the desired end is now reached may, and doubtless will, be changed greatly and in many ways; for it would ill become one to say that the simplest and best modes of reducing these principles to practice had yet been attained, and that future study and future labour would yield no further improvements. But of one thing I am persuaded, and that is, that we owe to Pasteur and Lister a very great addition to the means at our command for prolonging life and preventing suffering, which, I need hardly add, are the legitimate aims of all true surgery.

The antiseptic or Lister's system owes much to its opponents—more, perhaps, than to its friends. It is safe to say that had not its every part been challenged, its every advance rigorously criticised, it would hardly occupy the place which it does to-day. In science, as in literature and in politics, fair and searching criticism exposes error and advances truth. Certainly no doctrine in all surgery has been more mercilessly handled or contemptuously treated, more sneered or laughed at, than this of Lister's. But with truth and logical scientific deduction as its foundation, the waves of opposition which beat against it fortunately serve to show a weak point in its structure of minor importance, here and there, while the solidity of the principle on which it rests remains conspicuously undisturbed.

When seen from Lister's point of view, the elements of surgery on which the practice rests are, in a certain sense, changed. Where the principles are founded in truth, they of course, remain unaltered ; but where errors obtained, a better light has enabled us at least to detect their presence ; and while we still continue ignorant on many points, the consciousness of that ignorance is clearly a step towards its ultimate removal. Numerous difficulties have already been cleared up, knowledge has taken the place of obscurity, and complexity has yielded to simplicity. Much yet remains to be done.

I wish here to state that while I know no words in which to express what I owe to the teaching and example of Mr Lister, yet I alone am responsible for the statements I may make in these lectures—statements which I ask may be regarded as simply the expression of opinions formed during my labours in the Edinburgh Royal Infirmary, which begun as house-surgeon under Syme, were continued as assistant surgeon to Mr Spence and Mr Lister.

The Study of Principles.—The derivation of the term surgery, or chirurgery, indicates very clearly that at one time it was looked upon as a "handiwork;" and those diseases in which manual means were used to obtain a cure were placed under the charge of chirurgeons or surgeons. But from being merely a practical art, surgery has developed greatly in modern times, so that we have now the two well-marked divisions of *principles* and *practice*. In this course of lectures I shall speak of the Principles of Surgery ; and it is not intended to treat of the Practice, except as illustrative of the main subject. Let me demonstrate what I mean. In the application of a splint to a fractured limb, the principle is that we should command the break above and below the seat of fracture ; and from this arise many variations in the methods adopted and in the materials employed. Wood, pasteboard, pillows, etc., are used according to the necessities of each case. Again, the principles of bandaging are to put the bandage on so that it won't come off, and to make the pressure equable. Different means and materials attain this end in different parts of the body. A third example is the importance of rest and the avoidance of unrest. In the healing of a wound the parts must be kept quiet, and the materials employed for making the stitches must be chosen according to their fitness for this purpose. It can not be right that you should apply indifferently a flexible silk stitch and a rigid silver stitch ; or that the stitches may, as you please, be either few and far between, where each has a large area to keep at rest, or many where each has an easy task to perform. The means adopted for securing rest for the various organs of the body when diseased are, in practice, as various as their functions. The eye is kept at rest by confining the patient in a dark room ; the brain by prohibiting reading and thinking as much as possible ; the kidney by employing other organs of the body to perform their functions, such as the skin by the use of diaphoretics ; and so forth.

But while we must always bear in mind that we speak here of Principles chiefly, it must be remembered that it is not always possible to refer to Principles. In many cases we can only be empiricists ; and it is well that it should be so. This only means that our knowledge has its limits, and that there are still principles to be discovered and truths to be learned ; and in surgery, as in everything else, that part of the science where the search for truth is still going on, and our ignorance is most manifest, possesses the greatest charm for the inquirer. I shall always take the opportunity of pointing out the gaps in our knowlege, which remain to be filled up by future investigators

Health and Disease.—The subject which surgery has for consideration is disease ; and without spending time in considering disease from a surgical, as distinguished from a medical, point of view, let us merely ask the question, What is disease ? The simplest reply is, a departure from health. This brings forward the question, What is health?—and to answer it, I shall make use of a figurative mode of explanation.

Health, as a standard, is the normal condition of the human body, which may be represented by a curve, varying in direction at different periods of life, and in different individuals at the same period of life. The ascending part of the curve represents the time of growth and development, from conception to manhood; a more or less level portion then represents the period of maintenance; and this is followed by a portion downward in direction, signifying old age and decay, ending in death. During the first period we have the deposit in the tissues in excess of the removal; and we have, on the one hand, an increase in the bulk of the organism, which is growth, and, on the other, an increasing complexity in the performance of functions, development. During the second period we have maintenance, and the deposit in the tissues is equal to the removal. During the third period—that of decay —the removal from the tissues is in excess of the deposit.

Taking this curve as our standard, we may figure to ourselves disease as a departure from it in a downward direction. It is the result of a displacement of the same forces which are at work in a state of health, and this misplacement and fall may take place at any part in the curve. A child born of syphilitic parents is weak and puny, and the passage downward to death may occur shortly after the curve is begun; it may be even before birth in a miscarriage. This departure may be very gradual, and may itself take the form of a curve more or less sharp, as in the various forms of illness; or it may be a sudden instantaneous fall, as in a fatal accident.

But there is resident in the body a force that struggles against this misplacement—a power of recovery. You may be injured by the passage of a cart-wheel over your leg, fracturing the tibia and crushing the tissues around : but recovery takes place. In the injured tissues growth and development set in, and from the point to which the curve has fallen a renewed upgrowth takes place. The power of recovery may not be able to lift the patient up to his former condition, and the rest of life will be spent on a lower level than the normal one. For example, after an ulcer is healed we have a scar or cicatrix. The cicatrix is not true skin—it is but an imperfect substitute for it; but this is the point up to which the power of recovery leads the injured tissue.

Action of an Irritant.—The misplacement of the vital forces in disease is termed an *injury*, using this word in its most general sense. An injury is suffered equally when the leg is crushed by a cart-wheel and when pneumonia is caused by a draught of cold air. The injury in these and all other cases is the result of the application of an *irritant*. There may be direct action of the irritant; this we had in the first example, and in all other similar instances, such as wounds, burns, etc.; or there may be indirect action of the irritant; the pneumonia is caused by such; or a swollen testicle may be produced by gonorrhea in this way. The irritants are the cart-wheel, the hot body, the cold draught, the gonorrhea; the injuries are the crushed limb, the burn, the pneumonia, the swollen testicle. Following the application of an irritant, there is generally a certain definite series of changes to be noted. The injury has a life of its own, and the phenomena in this life appear, grow and develop, are maintained and decay. When my hand is burned, the injured part may either get better or it may die, and this upward or downward progress is represented by a curve such as we have spoken of. When the injury is merely local the recovery or death will be local; but it is not always possible to confine the action of the irritant to a limited area, and then the recovery or death will be general; or the area of the injured part may be limited enough, but the part itself be an organ necessary to existence, and in this case also the effects will be general. Observe also that as before mentioned the recovery may be complete or partial.

The Physiology of an Injured Part.—We have now to consider more in detail the physiology of an injured part. There is however one exceptional case which does not admit of consideration in this way. The injury may be so severe and affect so powerfully the vital functions of the tissues, that death of the tissue may take place directly. The passage from life to death is immediate. Putting aside this one exception we note that a vital process goes on in the tissues after every injury, ending either in recovery or in death. During health there is a constant interchange of materials going on in the tissues. The blood is the agent which is continually occupied in conveying something to the tissues and in receiving something from the tissues. After the application of an irritant the blood still continues its functions, but there is no longer maintained the equilibrium between the two processes of addition and withdrawal. There is now an *increased afflux to,* and an *increased deposit in,* the injured tissue. The materials are there for the maintenance of the vital functions : there is however a misplacement of these materials ; they are carried to the injured tissue in excessive quantity, and they are deposited in the injured tissue in excessive amount. In health the functions of the blood are performed imperceptibly, and the functions of the tissue are maintained intact; in disease there is an appreciable alteration in the structure and disorder in the functional activity of the affected part.

I have mentioned incidentally that an irritant may act directly or indirectly. A swollen testicle may be due to a kick—an instance of direct action ; or it may be due to gonorrhea—an instance of indirect action. In indirect action we have the nervous system acting as a communicating channel for the irritation ; in direct action the effect is produced apart from the nervous mechanism. A beautiful experiment of Mr Lister's illustrates the direct action of an irritant. When the ciliated cells are removed from any part of the body where they exist, the cilia continue their usual lashing motion for *some time* after removal. Mr Lister removed a piece of the tongue of the frog, and observed with the miscroscope the ciliary movement going on. He then brought a hot iron wire near the cilia, and the movements were quickened. This was withdrawn, and again the speed of the movement fell. The hot iron was again brought near, and quickening was again produced, though not to such a marked extent. It was brought nearer still, and the motion now became slower and at last ceased. When the hot wire was suddenly brought near the stage of microscope the ciliary movement stopped at once ; the ciliated cell was dead. All the effects of the irritant here must have been produced without the intervention of a nervous mechanism, as there was none present. Thus we are able to say with certainty that an irritant may act directly on the tissues.

The foregoing experiment brings out other facts of very great importance. The first effect of an irritant, if not excessively powerful, is to stimulate the part and quicken the movements. If continued longer or if rendered more powerful the irritant depresses the tissues and the movements become slower. And if sufficiently prolonged and sufficiently severe the irritant may cause the death of the tissue. Illustrations of this are constantly occurring. Rub the back of your hand briskly and you stimulate the tissue and it becomes red. Apply a stronger irritant in the shape of a mustard poultice and you produce a blister, in which the vitality is depressed. With a severe irritant, as a red hot iron, you destroy altogether the life of the tissue. But the strength of the irritant is not the sole condition which determines its effect. A second condition on which the action depends is the strength of the tissue. On any one whose work requires long periods of sitting, the continued pressure on the gluteal region produces no injurious effect ; but let the same individual be confined to his bed for sometime with illness, so that the vitality of his whole body is at a lower level, and under these circumstances the pressure that was formerly borne with ease can no longer be endured. Bed sores form on the gluteal region : in other words the irritant which formerly produced hardly any effect now depresses and kills the tissue. Again, it is an unsafe thing to apply a mustard poultice to a child who is suffering from a severe attack of measles, because that which formerly acted merely as a depressant may now, in the enfeebled state of the body, go much further in its effect, and a sloughing ulcer may be formed ; that is, the tissue may be killed. The action of an irritant depends then, first, on the strength of the irritant ; and secondly on the strength of the tissue.

Such is the action of irritants and their mode of causing an injury. There are three elements in every part of the body which come into prominence at any part which may have been injured. These are the *blood vessels*, the *blood*, and the *tissues* surrounding the vessels. The study of the physiology of these elements is of the utmost importance if we wish to understand the pathological changes that take place in them.

The *blood vessels* which come into prominence in this consideration are chiefly the capillaries and small arteries. The walls of these are thin and membranous, and are probably a continuation of the epithelial lining of the larger arteries. The feature we have especially to note about them is that they allow of free outflow and influx of some of the elements in the blood and surrounding tissues.

The *blood* consists of two elements, the liquor sanguinis or blood-plasma and the corpuscles. The corpuscles are of two kinds, the yellow and the white. The white are much the less numerous. They are more or less globular in shape, and possess a nucleus. They display amoeboid movements, shooting out and withdrawing processes of their protoplasm, and by this means moving about from place to place. They absorb substances from without, and excrete effete matters. Growth, development, maintenance, and decay go on in each of them individually. Thus we are able to predict that their function will involve active movement. The yellow corpuscles (red when seen in mass with the naked eye) are much more numerous than the white corpuscles, but perform a merely passive function, carrying oxygen to the capillaries where it is appropriated by the tissues, and carrying back the effete carbonic acid which is excreted by the tissues. In the ordinary state of the blood the corpuscles float freely in the liquor sanguinis, but there is a remarkable change which the blood frequently undergoes and the conditions and nature of this change are of special importance in a pathological respect. When blood flows from a wound it clots or coagulates. The importance of this change may be imagined when we hear John Hunter saying that there is more to be learned of the use of the blood in the animal economy from its coagulation than from its fluidity. Hunter held that wounds may heal by blood-clot, and of this we shall have a good deal to say hereafter. In the meantime let us direct special attention to the nature of coagulation.

For a long time it was held as an unquestioned truth that the coagulation was simply the change attending the death of the blood. But I trust to be able to show you that it is anything rather than a process of death. Several distinct steps are observed in clotting. The first change we observe when blood is poured out of the living body into a vessel is, that it grows viscid and flows with difficulty. It then passes into the form of jelly or clot. If you observe the surface of this jelly narrowly you will see globules of a watery-looking fluid gathering on it. If the vessel be transparent it will also be noted that a layer of this watery fluid lines the sides of the vessel, and that in fact the clot has now contracted greatly and is suspended in the fluid. If this clot be washed the red corpuscles will be carried away. What is left is a white stringy material, fibrin. This fibrin does not exist in the fluid blood. It is formed by a chemical union of two substances—fibrinogen and fibrinoplastin. The fibrinoplastin is in the white blood corpuscles, and the fibrogen is in the blood plasma. A third element is referred to by some observers—the fibrinferment— which is supposed to set up the change. But this we are really ignorant of as yet. The watery fluid in which the clot is suspended is serum or the liquor sanguinis after the fibrinogen has been substracted from it to form the clot.

A modification of the process is observed when the coagulation takes place very slowly. The coloured corpuscles form rouleaux, and when the clotting takes place quickly these rouleaux become entangled in the meshes of the fibrin that form : but if the formation of fibrin is from any cause delayed, or takes place very slowly, these rouleaux have time to sink to the bottom, which they do readily enough. We then have a clot which is colourless. This may be produced by cold or the addition of alkalies to the blood. And just as these reagents produce a colourless blood-clot by delaying the

formation of fibrin, so on the other hand it may be produced
by increasing the size only of the rouleaux. As long as the
coloured corpuscles float singly or in small rouleaux, they will
not sink quickly, just as one feather may float for a time in
the air ; but just as a bundle of feathers comes down immediately, so does the large roleaux of coloured corpuscles sink
rapidly. The formation of rouleaux seems to be hastened,
and they collect in large masses in the blood of an inflamed
part, possibly by rendering the corpuscles stickier ; and hence
we have the appearance known as the " buffy-coat "—a layer
of colourless clot on the surface of the coloured clot. The
" cupped " appearance of the clot is most marked when a
buffy coat is present ; in it the fibrin is increased relatively to
the blood corpuscles. The contraction is due to the fibrin ;
hence, the greater cupping seen when the blood corpuscles
are in small amount.

What is the cause of coagulation? We have coagulation
when the blood flows into an ordinary vessel. We frequently
have coagulation within the blood vessels when they are
injured ; and we have clotting between the lips of a wound in
which the tissues are injured—the injured tissues are depressed,
their vitality is lowered in consequence of the irritation.

An aneurism may be cured by the formation of a blood
clot in the sac : the walls of the cavity are in a condition of
lowered vitality. If a needle be pushed through the wall of
a living vessel through which the blood is flowing, and then
be examined after some time, a clot is found to have formed
around it. From facts such as these the conclusion has been
arrived at, that clotting takes place when the blood comes into
contact with dead or dying matter, or with tissues which
by the application of an irritant have been depressed: their
vitality has been lowered. An example of dead matter will
be afforded by the jar into which the blood is poured, or by the
needle passed into the cavity of a living vessel. An example
of depressed tissue will be found in the wall of the aneurismal
sac, or in the lips of a wound which are in a depressed state
of vitality, and to that extent are dying—are an approach
towards death.

In the *tissues*, the cellular elements are those which more particularly claim notice from us as surgical pathologists. And when I speak of the cells, I do not take the word in any one of the numerous meanings which have been attached to it. Whether we take it as meaning a neucleated mass of protoplasm with a cell wall, as Schwan would have us do ; or whether we discard the cell wall, with Max Schultze ; or whether we hold with Stricker, that a simple mass of protoplasm, without nucleus or boundary wall, is the cell—matters not. That thing which is the active agent in the tissues, which takes up nutriment and elaborates it to form muscle, bone, etc., is what we understand as the " cell." The body consists of a mass of innumerable cells ; and the sum of the life of these is the life of our bodies. Each of them is produced from parents,—grows, develops, is maintained and decays, and can act, to a certain extent, an independent part when occasion requires it.

The most convenient way of examining the tissues and the circulation, in a state of life and health, is to do so in some transparent membrane which may be suitable for microscopic purposes ; and such a membrane is ready to hand for us in the web of a frog's foot, or in the mesentery of the frog. On directing attention to a particular vessel we observe the current of blood hurrying on swiftly in the centre of the tube, but at the sides there is perceptibly much slower movement. Here the colourless corpuscles are seen moving along very slowly, and now and then sticking for a time to the wall of the vessel. In such a membrane we can observe the pathological changes which follow the application of an irritant ; and this brings us to a consideration of the phenomena to be observed in an injured part, *i.e.* an irritated part. In the web of the foot irritation may be produced by the application of heat, or by such agents as mustard or chloroform. If we choose the mesentery of the frog for observation, the exposure to the cold air is all that is necessary for irritation. Attention is now to be directed *(a)* to the vessels, and *(b)* to the blood flow.

In the *vessels*, the first thing to be observed is contraction, which, however, is very evanescent, and has often not been observed at all. Following closely on this there is *dilatation* of the wall, and increase in the caliber of the vessel.

The *rate of the blood flow* is, in the first instance, slightly quickened. This, like the contraction of the vessels is merely momentary, and is quickly followed by a slowing of the current, so that you can observe the size and shape of the corpuscles quite easily. This slowing goes on increasing until complete stoppage or *stasis of the blood flow takes place.* This stasis and the dilatation of the vessels are the essential result of irritation. Their causes are exceedingly obscure, and we can not speak with certainty on any point here involved. But still something has been done of late years to clear up the difficulty; and I will now briefly communicate the results of recent researches.

First, as to dilatation of the vessels. This follows on irritation ; and you will remember that I said before that irritants act in two ways, directly and indirectly—directly on the tissue to which they are applied, and indirectly through the nervous system; and that the effect of an irritant, although at first stimulant, is mainly depressant. In the subcutaneous tissue of the frog, there are spaces filled with pigment, which communicates a colour to the skin. In a state of health the frog is light in colour, but when the frog is out of health the colour darkens greatly. Microscopic observation reveals that during health these pigment spaces are small and compact—the frog is of a light colour ; but that during an unhealthy state, they expand greatly by shooting out processes—the frog is dark in colour. When the web of a healthy frog is irritated, the pigment spaces expand in the irritated area; in the surrounding parts they remain in their normal contracted condition. We can see the effects of the irritant on the tissue composing the pigment space ; and we are justified in assuming that the irritant acts in a similar way on the walls of the blood vessels. They also lose their power of contraction; their tissue is also depressed or injured, and the injury is the result of the

depressant action of the irritant. The evanescent contraction
sometimes observed in the early stage is due to the stimulant
effect of the irritant. The irritant acts directly on the tissue
to which it is applied.

Somewhat different is the case of the irritant which acts
indirectly through the nervous system, as, for example, when
an attack of pneumonia follows exposure to a draught of cold
air. It has been shown that if the sensory nerve of the ear
of the rabbit is stimulated, this stimulation is followed by
dilatation of the vessels of the ear ; this dilatation is due to a
change in the condition of the vaso-motor centre from which
the nerves which supply the walls of the vessels take their
origin. This change in the vaso-motor centre is one of depression ; it loses its command over the vascular walls—the
blood vessel dilates. We have, I believe, in this experiment
the clue to the explanation of the way in which the indirect
irritant acts. The sensory nerves of the part to which the
irritant is applied convey an impression to the vaso-motor
centre ; it is depressed, and as a consequence the vessel dilates.
There is undoubtedly a nervous connection between the skin
over an organ and the organ beneath—between the skin of
the chest and the lung; there is also an intimate nervous connection between organs in functional relation, as the ovary
and mamma, the urethra and testicle—during ovarian irritation the mamma is turgescent, during an attack of gonorrhea
the testicle may inflame.

Secondly, the stasis requires explanation. And I now ask
you to go back to that part of these lectures where I spoke
of the coagulation of the blood. Remember that I tried to
impress upon you that there were three stages in clotting—
viscidity, jellying and contraction. We also saw reason to
believe that these results followed when blood was brought in
contact with depressed tissue. The dilated vessel indicates
depression. The blood in contact with the dilated vessel becomes sticky or viscid ; this viscidity first slows its flow, and
if the depression continues, the viscidity increases, the blood
flow stops—stasis takes place. The stasis only occurs in the

irritated area, and at the edges of the stasis area, where the
blood current is still going on, the stasis blood is constantly
being washed away by the fluid blood. This was shown first
by Lister, and he first drew the deduction that the statis is
due to the condition of the vessel walls.

In consequence of the dilatation of the vessels in the irritated
area we have an increased quantity of blood in and necessarily
an increased afflux to the injured tissue. It remains now to
consider how there is an *increased deposit* accompanied by an
increased production in the injured part.

The increased deposit comes from the blood. As you are
all aware during healthy nutrition, blood plasma is constantly
flowing through the walls of the capillaries. It is poured out
in excess in an injured part, as you may observe from the fact
that an injured muscle is softer than a healthy one. If you
observe also the dressing that is taking off an incised wound,
you will see that it is saturated with a clear fluid. This exudation is the blood plasma. There is increased intra-vascular
pressure in consequence of the stasis in and increased afflux
of blood to the injured part : as a result there is increased
exudation of the fluid portions of the blood into the irritated
tissues through the dilated vascular walls. The thinning of
the vascular walls is also an important factor in the process.
The fluid portions of the blood pass more easily. Of the two
kinds of corpuscles I see no reason to believe that the coloured
perform any essential part in an irritated tissue. They are
certainly to be found outside the vessels, but the rupture of
the thin walls of the dilated vessel will account for this. It is
different with the white corpuscles. Williams long ago pointed
out the tendency of these bodies to cling to the sides of the
vessel. Addison in 1842 maintained that pus corpuscles and
white blood corpuscles were identical, and that the latter
migrated through the walls of the capillaries : and though the
hypothesis was strengthened by Waller, who in 1846 said that
he had seen the migration going on, yet the statements of these
observers were unheeded for many years, until the same story
came from Germany ; after Recklinghausen in 1863 published

the results of his study of the white corpuscles, and proved that they had a capacity for movement in themselves. He directed attention to what had been previously shown by Wharton Jones in 1846. He observed them shoot out and withdraw their protoplasmic processes, and by treatment with colouring matter he discovered that they had the powers of absorption and excretion. In 1868 Cohnheim demonstrated that when a part is irritated the colourless corpuscles collect in great numbers, and that by means of their processes they can make their way through the membranous walls of the capillaries. This process takes place normally in the healthy tissue, though it requires great patience to see it; but it can be seen without difficulty in an irritated part. I have seen these corpuscles pass through, and after their passage move about in the surrounding tissues and there divide.

But this is not all: in addition to the increased deposit of the fluid and corpuscular elements of the blood, there is increased production in the irritated tissue. This increase is due in part to division of the migratory white blood corpuscles, but in an irritated part there is cell proliferation also. The cellular elements proper to the tissues multiply by rapid division. This is due, I believe, to overfeeding of the tissue. This proliferation goes on in health as the normal method of growth, but in disease it is in excess. It must not be supposed that the increased production is an evidence of increased vitality. The increased proliferation is due to the overfeeding of the tissue ; the normal equilibrium is lost, the cell elements are overfed, they proliferate, their normal functions are in abeyance like a hot house vine highly manured, which brings forth many bunches which if allowed to remain will never come to maturity, like the stunted trees in a wood which are never thinned ; in both there is an increase in the number of the bunches and plants at the expense of the parent vine and the neighbouring trees. With an increased growth we have decreased vitality, a tendency to degeneration and decay; hence the numerous cell elements in an irritated tissue do not go to form normal tissue, but form various aborted products, the chief of which is pus. It is at present an unsettled question as to the chief source of the pus. Is it due mainly to proliferation of the migratory white blood corpuscles, or to proliferation of the original cell elements of the part? The truth probably is, that both causes are at work in highly vascular tissues; in nonvascular parts, as shown by Goodsir, in cartilage the original cell elements are the main if not the whole source of pus supply.

The formation of "Inflammatory Lymph."—When these changes have taken place, and the balance of nutrition has been destroyed, there is produced a substance which to some may appear altogether of a peculiar nature. There are present in the irritated tissue the liquour sanguinis, which has been poured forth in excess, and the white corpuscles, which have found their way through the walls of the vessels. In these two substances have we not all the elements for the formation of the fibrin of a blood clot? The irritated and depressed tissue is there *ex hypothesi;* the fibrinoplastin and fibrinogen will, in an ordinary case of irritation without rupture of the vessels, be unencumbered by the presence of the yellow corpuscles. All the circumstances favour the formation of the colourless blood clot. But you will not find that the identity of the substance that forms in an irritated part with a colourless clot is recognized in the name that is given to the former. "Inflammatory lymph," the name referred to, is an objectionable expression, for this reason, that it multiplies distinctions where none really exist. It prevents us seeing the application of principles of sound physiology to the treatment and examination of pathological questions. And it is because I am so desirous that you should clear your mind from the confusion of ideas that is sure to follow the using of different terms for the same thing, that I have been thus minute in the treatment of the physiology of the blood. And henceforth you will understand that I use the terms "inflammatory lymph" and "colourless blood clot" as synonymous.

Process of Recovery.—At several stages in the preceding series of events a stop may be put to the process, if the irritant is withdrawn. In simple language "recovery" may take place. When there is active congestion and dilatation of the vessels, withdrawal of the irritant produces recovery of tone, and at the venous extremities of the capillaries the corpuscles commence to break away and resume their normal course—a process which soon spreads over the whole affected part. If there has been stasis of the flow, the way of recovery is the same. If there has been effusion of the liquor sanguinis and white blood corpuscles into the introcellular spaces, these are again absorbed by the vessels and resume their normal functions. But if "inflammatory lymph" has formed and occupies a definite area, there is the possibility of a new mode of recover; and the clear comprehension of this last is of the utmost importance in the Principles of Surgery. All the injuries which call for surgical interference present examples of it, and accordingly the whole question of the *Process of Repair* now opens out before us.

THE AMERICAN PRACTITIONER.

JUNE, 1879.

Certainly it is excellent discipline for an author to feel that he must say all that he has to say in the fewest possible words, or his reader is sure to skip them; and in the plainest possible words, or his reader will certainly misunderstand them Generally, also, a downright fact may be told in a plain way; and we want downright facts at present more than anything else.—Ruskin .

Original Communications.

THE ELEMENTS OF SURGERY.

BY JOHN CHEINE, M.D., F.R.C.S.E.

Surgeon to the Edinburgh Royal Infirmary, etc., etc.

LECTURE II.—Hemorrhage—Structure of the Vessels—Phenomena of Hemorrhage and its Natural Arrest—Causes of these Phenomena—Bleeding from an Injured Tissue—The Method of "Bloodless Surgery"—Artificial Arrest of Hemorrhage—Torsion—Ligatures — Plugging — Acupressure — Styptics—Caustics—Varieties of Hemorrhage—Primary Hemorrhage—Reactionary Hemorrhage—Secondary Hemorrhage—Effects of Severe Hemorrhage.

THE PROCESS OF REPAIR: First, in a Wounded Vessel; Second, in the Injured Tissues.

Hemorrhage.—With my remarks on the process of repair, I intend to associate very closely what I have to say on the subject of hemorrhage. The irritation which is accompanied by laceration of the tissues presents at first sight a much more complicated series of phenomena than that which I have been dwelling upon. The process of exudation does not go on alone. Each tissue in the body, with one or two exceptions, is practically a sponge traversed in all directions by minute canals filled with blood; and any break in the continuity of a vascular part is accompanied by rupture or wounding of some of these canals. Hence we have hemorrhage.

A short consideration of the process by which a wounded vessel is repaired will, in my opinion, be the best introduction to a consideration of the process of repair in the tissues generally. The wounded vessel heals by the formation and organization of blood clot. I shall try and show that the tissues are also repaired by the formation and organization of blood clot—a doctrine first taught by John Hunter.

There are, as you will know from your physiological studies, certain differences in the flow of blood from a wound, according as the blood comes from an artery, a capillary, or a vein. From an artery we have a succession of pulsatile jets of bright red blood; from a vein we have a steady flow of dark-coloured blood; and from a capillary, or rather from capillaries, we have a slow oozing of blood, intermediate in colour between venous and arterial blood.

Structure of the Vessels.—To understand what takes place when these vessels are cut across, recall to mind what you know of the anatomy of an artery, a vein, and a capillary. In an artery there may be recognised for our purpose three coats. We have lining the vessel, and in immediate contact with the blood, an epithelial coat modified in function by the presence of some elastic tissue. This is very fragile and easily torn, cut or destroyed. Secondly, we have the middle coat, much thicker, and consisting of elastic tissue and non-striped muscular fibre—the muscular coat. The arrangement of the fibres is circular, and this must be particularly noted for two reasons, because it renders this coat of the artery weak to a force applied circularly parallel to the direction of the fibres, as for example when we apply a ligature; and also because it produces a tendency to constriction of the cut end of the vessel. Thirdly, there is an outer coat—the elastic coat—consisting of oblique and longitudinal fibres. The distinguishing peculiarities of this coat also arise from the direction of the fibres. It is strong to a force applied circularly, as in the case of a ligature, and there is a tendency in its fibres to retract when the vessel is cut across. I here show you a portion of the femoral artery, which will illustrate part of what I have said. It has been ligatured, and you will note that of the three coats the inner and middle have given way, while the outer still retains its continuity: the inner and middle coats—the former in consequence of its weakness, the latter in consequence of the circular direction of the fibres—have been divided with the ligaturè as with a knife. Besides these coats, properly so called, it is necessary to take account of the sheath of the vessel, consisting of areolar tissue, and *loosely* connected with the external coat. The loose connection of the sheath with the vessel allows an artery to retract within its sheath when it is cut across.

A vein differs in structure from an artery in two particulars; it is less muscular and it is less rigid. Hence, there is greater tendency to collapse, and less tendency to retraction and contraction, when the vessel is torn or cut across. But it is powerfully elastic, and will bear a ligature.

The capillary is practically the continuation of the internal epithelial coat of the artery, and tends to collapse when lacerated, being impelled to this besides by the elasticity of the surrounding tissues.

Phenomena of Hemorrhage and its Natural Arrest.—With these facts in mind, let us direct our attention to an open wound in which no large artery is cut. At first there is seen a general bleeding from the whole surface. It is an oozing from the capillaries and small arteries. This after a time ceases, and the flow is restricted to (in a small wound, perhaps) one or two spots, where we observe pulsatory jets of bright red blood. Gradually this jet becomes less in size; it issues forth with less force, and is thrown to a less distance, until at last the stream comes away drop by drop, and ultimately a clot forms arresting the flow altogether. This is the *natural arrest of hemorrhage.*

Causes of these Phenomena.—I have now to ask your attention to the causes of the foregoing phenomena. The great thing needful in the matter. I may begin by saying, is the coagulation of the blood. There is a peculiar constitutional state found in some people to whom the preceding description does not apply in cases of bleeding. They are the subject of what is called the " hemorrhagic diathesis," or " hemophilia." In individuals subject to this condition, it is very difficult, sometimes impossible, to stop bleeding. The slightest wound, such as that caused by the drawing of a tooth, may cause persistent bleeding even to death, and pressure, styptics, cauteries, etc., are all tried in vain. As to the nature of this diathesis we are in ignorance. There can be no doubt that the thorough investigation of the subject would reveal some unknown conditions which would render cases of the hemorrhagic diathesis more amenable to treatment. Any light thrown on the true pathology of this peculiar condition would also add to our knowledge of the causes of the coagulation of the blood.

To return to the normal condition. In the case of any solution of continuity, you will observe that in the capillaries and arterioles there is collapse of the walls, and with the irritation at the lacerated or cut end of the vessels the coagulation of the blood and stoppage of the blood flow very soon supervenes. With the large arteries in which the three coats are developed (I do not speak of the very large arteries in which the rush of blood is so great as to prevent any natural arrest), the process is more complicated. The muscular fibres of the middle coat, and the elastic fibres of the outer coat, contract simultaneously. There is a constriction of the lumen or opening of the vessel, and there is retraction within the sheath which collapses and assists to make the opening for the blood flow still smaller. The time that the following changes take to occur, will depend on the size of the vessel and various other extraneous conditions. We notice, in the first place, a coagulum forming on the sides of the channel lined by the rough sheath of the vessel. This coagulum gradually increases in size until the channel is completely closed, and the bleeding is arrested temporarily. The coagulum then extends upward in the lumen of the cut vessel for a variable distance, generally to the first branch ; it can not extend further in a proximal direction, its formation being prevented by the current of blood constantly washing over the apex of the clot. The external clot acts as a temporary obstacle to the fl ow of blood. The internal clot, which is attached by its base to the external clot, protects the external clot from the force of the current, and allows those changes to take place in it which ultimately end in its organization ; the result is the permanent closure of the wounded vessel.

The tissue which is now stopping the escape of the blood is a blood clot—recent, rudimentary, and by no means strong. At any moment the current may receive an access of strength, and sweep the temporary obstruction away. During the time necessary for the formation of the clot the individual will no doubt have lost blood, and a weakened circulation is the result. This favours the chances of the current being completely stopped ; but the moment that this is attained, and no more blood escapes, the circulation will commence to recover its tone. In other words, the clot is formed under a lower pressure than that which follows shortly after the complete closure. Further changes then are needed, and in the clot there is formed a much stronger tissue than the mere fibrin of coagulated blood.

The series of events which take place, ending in the organization of a blood clot, have been carefully traced. The clot is formed in the sheath from which the vessel has retracted, and in the vessel itself as far as the first branch. It now becomes *adherent*, in whole or in part, to the walls of these structures. At the point where the external clot in the sheath joins the internal clot in the vessel, you will next observe an alteration in colour. It assumes a *lighter hue*. This gradually extends throughout the whole mass. Microscopic examination reveals that it is due to two things. There is a disintegration and removal of the coloured corpuscles: their function is gone, and they now form simply an impediment to the process of organisation; therefore they are got rid of. But there is also a great increase in the cellular elements of the clot.

Migration of the white blood corpuscules takes place, and it is generally believed that in such situations there is rapid proliferation of these elements. The connective tissue corpuscles of the surrounding structures, more especially the endothelium of the inner coat, also proliferate and their products pass into the new tissue. The third step is *vascularization.* Blood vessels shoot into the clot. These are chiefly from the vasa vasorum, but it seems probable that there is direct communication established with the lumen of the vessel that has been occluded. This vascularity is only a temporary state of the clot, and but the means for establishing the ultimate condition of things. Gradual *contraction* is noted as the fourth stage in the process. This occurs after the tissue has become vascular ; and, *pari passu* with the contraction, there is noted a decrease in the vascularity of the tissue, due to a disappearance of many of the new formed vessels. The intercellular substance in the clot becomes fibrous, the cell elements are less evident, perhaps they elongate and form fibres, and gradually the whole clot is transformed into fibrous tissue ; and if we examine carefully the tissue which occupies the situation of the primary blood clot, we will only find in its place a fibrous cord, which gradually tapers off into the tissues. This is the last stage in the process. Such is the natural history of the permanent arrest of hemorrhage.

The preceding remarks have reference to an artery. The changes are slightly modified in the case of a vein. There is less tendency to contraction and retraction, but much greater to collapse : and the fact that there is no pulsating strong current also contributes to the ease with which coagulation takes place.

Such are the natural laws by which hemorrhage is conditioned, modified and suppressed ; and you will, perhaps, bear with me if I speak shortly of one or two practical applications.

Why does a wound *in* an artery bleed so furiously? This is illustrated by the operation of arteriotomy. This is performed on the temporal artery, and the noteworthy feature is that the surgeon having carefully dissected through the skin

and fascia, down to the vessel, makes an oblique opening in
it. The blood comes out with great force, and when enough
has been drawn, the vessel is cut across, a pad put on the
wound, and fixed in position with a bandage. When the first
incision in the vessel is made, the contraction and retraction
of the coats, if they take place at all, only assist in enlarging
the opening ; but when the artery is cut across, then the effects
which have been already dwelt upon follow at both ends of
the divided vessel; and these, assisted by a pad and bandage,
are sufficient to stop the bleeding.

Bleeding from an Inflamed Tissue.—We may also explain
in· part why it is that in an inflamed and conjested tissue we
have such copious bleeding. No doubt an inflamed tissue
contains more than the normal amount of blood ; but besides
that the contraction and retraction of the cut vessels do not
readily take place, as they are more firmly united by inflammatory adhesions to the adjoining tissues ; and this prevents
the speedy stoppage, while the congestion affords a larger
supply of blood. The artery cannot retract and contract as
in the healthy tissues.

The method of "Bloodless Surgery."—A third application of
these principles is of great importance. Esmarch's system
of "bloodless surgery" has been extensively employed of
late years, both in this and other schools. It consists in the
use of an elastic bandage to squeeze out all the blood
from a limb that is to be amputated or operated upon, and in
the substitution, in place of the tourniquet, of a strong piece
of elastic tubing wound tightly round the limb to prevent the
flow of blood into it during the operation. No doubt the
term "bloodless" * is justly applicable to this method *during*
the operation ; but in its very completeness to my mind there
is a serious defect. If the limb is emptied wholly of blood,
there is no material for the formation of clots. The channels

---

* I do not desire here to discuss the question whether or not it is a right thing
to leave the blood and take away the limb. It is apparently taken for granted by
the admiration surgeons have for "bloodless surgery." To my mind it is by no
means a settled question.

of the vessels are kept completely clear for the rush of blood when it comes ; whereas if the limb is not entirely bloodless we have, during the performance of the operation, gentle trickling of blood with little or no pressure—the most favourable condition for the natural arrest to take place. The elastic tubing cannot be slackened and again tightened ; and if any of you, gentlemen, should adopt it in your practice, I hope you will have first considered that you will have to deal, not with the advantages and aids which you have in the hospital theatre, but with the notorious poverty of resources of a private practice. Not only have you no assistance from the formation of blood clots; you have not even the usual minute streams to guide you to the smaller arteries before you unloose the tubing. So little can be done before the elastic tubing is taken off, and so much requires to be done immediately after it is taken off, that Prof. Esmarch, the inventor of the method, employs twenty-four pairs of forceps at one of the major operations. Can any one work with twenty-four pairs of forceps at a private operation, where perhaps he could not secure the services of a single qualified assistant ? Let me ask you, then, to consider whether the conditions of hemorrhage admit of the employment of this method anywhere but in a hospital theatre. It is, in my opinion, a better plan simply to raise the limb vertically for two or three minutes before applying the tourniquet. This renders the limb sufficiently bloodless for all practical purposes. A small amount of blood in the vessels is useful, because blood is necessary in order that natural arrest of hemorrhage may occur. The trickle of blood guides the surgeon to the vessels which require ligature. The tourniquet can then be slackened and again tightened, more bleeding points will be observed, and after they are ligatured the tourniquet can be taken off. By this method less blood will be lost in the long run, and the practitioner will be enabled, by the use of these means, to perform with little professional aid the major operations in surgery.

Artificial Arrest of Hemorrhage.—We have treated of the
arrest of hemorrhage in the capillaries, veins and small arteries. Take now the case in which a large artery has been
cut. If you dared to stand by and watch with a scientific
interest the appearances in this case, you would in a short
time observe a decrease in the magnitude and force of the
current. But mark the reason : an immense amount of blood
has been already lost ; there is no longer the same amount
of blood flowing into the ventricles at the diastole; at the
systole there is no longer, therefore, the same purchase gained
by the muscular fibres in their contraction, and the propelling
force of the heart's action is materially diminished. Besides
this the total amount of blood in the body is rapidly growing
smaller and smaller. In short the decrease in the magnitude
and force of the current is associated with failing vital power,
and the hemorrhage, if unchecked, will bring on fainting and
death. It is useless to hope for a natural arrest by anything
short of these two last issues : the waste is too great. The
surgeon's active aid is required, and artificial means are necessary to stop the bleeding. The means at command are
torsion, the ligature, plugging, acupressure, styptics, and the
cautery. These artificial means take the place of the external
clot in the natural current. They temporarily stop the flow
of blood, and allow time for the permanent arrest to take
place. It is needless to remark that the surgeon employs
these devices to stop the bleeding from vessels of any size,
as the process of natural arrest always involves the loss of a
certain amount of blood.

(1.) *Torsion.*—This is a process by which, it has now been
proved, we can stop the bleeding from any vessel, however
large. In order that it may be efficacious, you must first,
with your forceps, lay hold of the artery only—a matter of
some difficulty in a small vessel. Then draw it forward clear
from the surrounding tissues, and slowly twist it. A few
turns only are necessary, and when let go it remains twisted,
with the inner and middle coats ruptured, and the outer
remaining entire. A clot forms in a short time, in which, by

the process of organization, fibrous tissue is developed, as already described when a vessel is closed by natural arrest.

(2.) *Ligature.*—This is the commonest mode, and, everything considered, is, where it can be employed, the best mode. As it is only by actual practice that you can really know how to apply a ligature, I shall not detain you with the minutiæ. But I do wish to say a word with regard to the materials employed. The substance which has been most in favour with surgeons is silk. Now silk is practically a foreign body; and a foreign body present in the tissues causes a certain amount of irritation, which leads to ulceration and local death; and, as a matter of fact, the silk ligatures do generally come away in a small slough. There are cases in which this ligature is encapsuled with fibrous tissue, but they are rare. It is not difficult to see that a process of ulceration going on in a wound that ought to be healing, is fraught with a certain amount of danger, and the desire to avoid this danger has been the source of contrivances other than ligatures which have been hit upon. But Mr Lister, reviving a practice of Sir Astley Cooper's, has introduced catgut as the most suitable material, and this combines in a remarkable way all the necessary qualities. Its strength is nearly equal to that of silk; and it does not, as silk does, form the nucleus for an ulcer. *It is an animal membrane, and can be absorbed in an animal tissue.* The process is slow enough to allow of the formation of a firm clot before absorption takes place. But from the first the catgut is not a foreign body; it is, if I may so speak, welcomed by the tissues amidst which it is placed, and friendly relations are established at once.

Sometimes it may not be easy to lay hold of the vessel in order to apply a ligature; the readiest method in such a case is to pass a curved needle, threaded with catgut, through the tissues, under the vessel, and this ligature is tied, including the tissue, around the vessel along with the bleeding artery. This method is speedy and efficacious, and might with advantage be used more frequently. It has one great advantage that there is no chance of the ligature slipping; it is held in position by the tissues included in its grasp. In an amputation of the thigh, I lately used it to secure all the bleeding points, after tying the femoral artery and vein in the usual way. I was much pleased with the ease, speed, and security with which the bleeding was checked, and from what I saw in that case I intend to use it *systematically* in operations. The needle may be threaded with a piece of catgut of considerable length ; and in this way one needle will do for several vessels. In the case mentioned I required to use three needles, each with a thread eighteen inches long.*

(3.) *Plugging.*—This is adopted in those situations in which it is undesirable to make a large wound in order to render the ends of the vessels visible, as in the palm of the hand and sole of the foot. In these localities a large wound would form, when healed, a large cicatrix ; and the weak tissue of which a cicatrix consists could not undergo the amount of pressure which is demanded of the skin in these parts of the body. Plugging is also employed in cases of hemorrhage in any of the passages in the body where it is impossible to gain direct access, as in the rectum and vagina. The principle of the application of a plug is that it should be dry. If wet it will form practically a poultice, and therefore a source of more bleeding. It should also if possible be wedge-shaped, the apex of the wedge in contact with the bleeding point, so that advantage may be taken of this form to obtain firm pressure, as the application of a bandage used to bind down the plug. This latter remark applies chiefly to wounds in the palm and sole. The necessary pressure is afforded in cavities like the vagina and rectum, simply by the opposing walls.

(4.) *Acupressure.*—I only retain mention of this method in my lectures out of respect for the inventor; because the introduction of torsion, catgut ligatures, and antiseptic precautions, now secure every one of the advantages which were specially claimed for acupressure by the late Sir James Simpson.

---

* April 20, 1879. These words were written in January; at all my operations since that time the arteries have been tied in the way I mention. I have no reason to be dissatisfied with the result ; it is quicker and simpler than the old plan.

(5.) *Styptics.*—These are used in severe capillary hemorrhage. How some of them act is not very clear, but one way is by producing coagulation. This is the effect of a favourite styptic in this school—a mixture of perchloride of iron and glycerine. I may mention here that ten or twenty drops of the liquid extract of ergot, injected subcutaneously into any part of the body, will sometimes stop bleeding. How it acts we do not know with certainty: it is said to cause contraction of the non-voluntary muscular fibres.

(6.) *Cauteries.*—These are useful in those cases in which it is desirable that there should be the smallest possible loss of blood, in which the ligature is not available, and in which also the application of the tourniquet or elastic tubing is not possible. Operations in any of the cavities of the body supply examples. The cautery may only be used to arrest the bleeding, or the tissues may be divided with the cautery. The instrument, in the latter case, sears the vessels as it passes through the tissues.

Varieties of Hemorrhage.—I have hitherto spoken of hemorrhage quite irrespective of the agencies which have brought it about. But variations in the causes and circumstances call for variations in the modes of treatment. There are three great varieties of hemorrhage—primary, reactionary, and secondary.

(1.) *Primary hemorrhage* may be either (*a*) from wounds received by accident, or (*b*) from wounds made in the performance of operations. The first class of occurrences are those which claim our most serious attention. If you are called to a case where there is severe bleeding, immediately place your finger upon the bleeding point and exercise compression. If you can not do that, then compress the artery between the wound and the heart. The flow of blood having been temporarily stopped, proceed to take measures for the proper permanent arrest. Remove all objects about the wound, such as articles of dress, etc., which in any case are probably soaked with blood, and will, therefore, if allowed to remain, act as poultices. Even clots of blood will have this effect, and they too must be cleared away. The tourniquet having been ap-

C

plied, tie both ends of the divided artery at the bleeding point. The ligature below the bleeding point should never be omitted on account of the possible anastomosis bringing on bleeding in a short time. If you can not secure the artery at the wound, then tie it higher up between the heart and the solution of continuity. You will frequently be called to a case in which there has been severe bleeding, which has stopped before your arrival. If you can remain, you need do nothing ; but if you require to leave the patient, then open up the wound, find the wounded artery, and tie it. This is precautionary against reactionary hemorrhage.

In the treatment of wounds received at an operation, the circumstances of the case are all so defined and well known, that it is rather a question of the application of the simplest and commonest methods in the most expert manner than one in which we require the guidance of principles ; and this is work for the class on operative surgery rather than for these lectures.

(2.) *Reactionary Hemorrhage.*—This comes on within two days after the binding up of the wound, during the reaction which occurs when the circulation regains its power. The causes are chiefly instances of omission of some small arteries in ligaturing, not unfrequently the neglect to secure *both* cut ends of a bleeding artery. The use of Esmarch's tubing has also been shown to be a great cause of reactionary hemorrhage. The force used in applying it paralyzes the contractile power of the vessels, and they do not contract naturally. As to the symptoms, these are not at first well-marked. We have a stain on the dressing, small at first and increasing gradually. If there be much dressing, as there requires to be in the antiseptic system, then it may be very difficult indeed to detect this, and the first signs noticed sometimes are weakness and paleness from loss of blood. Whenever this hemorrhage is suspected the dressing is at once to be taken off, the wound opened up, and the mistake, if there has been any, remedied by tying the vessel. It will be found occasionally that clearing away the clot of blood that has formed is enough, as the poulticing effect of this is sufficient to produce and keep up the hemorrhage.

(3.) *Secondary hemorrhage* comes on after the first week, and may take place at any time until the wound is healed. Anything which interferes with the organisation of the clot is a cause. This interfering cause may be either a local one, as a putrefying state of the wound, or the special ulceration caused by silk ligatures, which may cause breaking down and destruction of the young growing tissue; or it may be a constitutional one, such as erysipelas or pyemia. The symptoms appear even more slowly and uncertainly than those of reactionary hemorrhage, but are otherwise the same, locally and constitutionally. The treatment consists in raising, if possible, the bleeding part, in order to take the blood-pressure off the wound as much as possible. The gentle application of a tourniquet may be used to effect this, and ice is used to induce contraction of the vessels at the part. I have seen very good results follow from the subcutaneous injection of ergot in the way I have already mentioned. If these means fail, you must break up all the new-formed adhesions, search for the bleeding point, and tie it. If the tissues are sloughing it may be impossible to tie the vessel in the usual manner, and then you must either use a threaded needle, include some of the surrounding tissue in your ligature, or try to sear the part with the cautery at a black heat. If you are still baffled, then ligature the artery of supply above the bleeding point.

Secondary hemorrhage is rarer now than it used to be. This is chiefly owing, I believe, to the fact that antiseptic precautions render the chances of ulceration much less, and the process of fibrillation in the blood clot goes on with much less hindrance.

Effects of Severe Hemorrhage.—Before leaving the subject of hemorrhage, let me interpolate a few remarks on the effects of severe hemorrhage. Now we see these only when a large artery is opened. Formerly, in the days of venesection, observation of the effects was much more frequent. The most marked result is the state called syncope. At the beginning of each session, you have in the hospital wards abundant opportunities for taking note of the phenomena attendant on this condition, for a mild form of fainting is very common with some of you on your entrance upon your practical duties. The pulse becomes weak and compressible ; the patient begins to yawn; pallor, muscular weakness and nausea quickly come on, followed by loss of vision, and he falls back in a "dead faint." The cause of all these symptoms is anemia of the brain, and on this fact is based the treatment ; for it is generally sufficient to lay the patient in a horizontal position and lift up his legs, and this explains also why syncope so seldom takes place when the patient is lying. In those severe cases caused by actual loss of blood from the body, we may have two results : First, there may be reaction, the patient become flushed, the respiration hurried, the pulse jerky, and there is singing in the ears. We do not interfere here ; but if there is a wound we must watch for hemorrhage, for the heart is now regaining its power of contraction. Second, however, there are those cases in which the patient's rallying powers are very slight. The pulse is small and intermittent, the intense paleness continues, the patient becomes cold with clammy sweats and passes into a dozing condition, and in this state may die. Urgent treatment is necessary. Stimulants are to be applied internally in the form of brandy, etc., and externally as (*e.g.*) mustard and hot bottles. Tourniquets may be applied to the limbs in order to keep the blood in the axial centres. The last device is merely taking blood from a part where it is not urgently required to supply a part which does require it. And as a final resource, when not even the limbs of the patient can supply the necessary amount of blood, it may be taken from the body of another. And here I wish to say a word about transfusion.

The first and chief requisite of any instrument for transfusion is *simplicity*. A basin half filled with warm water at one hundred degrees of Fahrenheit, a second smaller basin to be placed within the first for holding the liquid to be transfused, a canula to be inserted into the vein of the patient, and a syringe to hold eight or ten ounces of blood, are all that is necessary. The precautions of the operator are to be directed mainly against the chances of the entrance of air into the veins of the patient. A curious question has been raised as to the liquid that is to be injected. The first thought, of course, is that it ought to be blood simply as it comes from the body. But this in the process of transfusion necessarily comes into contact with foreign bodies, and acquires a tendency to clot, which of course impairs its efficiency. The introduction of clots into the circulation may give rise to dangerous complications. But the elements of fibrin are not required in this emergency, and Panum's experiments go to show that defibrinated blood (the fibrin being removed by smartly whipping up the blood as it is poured into the basin) supplies all that is necessary. But speculation has warrant in fact to enable it to go further. Is it the nutritive elements of the blood at all that is necessary here? Is not what is really wanted simply more force in the heart's contraction to enable it to send the blood to the head and vital organs? And will not this end be attained by the presence of any fluid in the heart on which the ventricles may gain a purchase, to enable them to contract? If any one has an opportunity of seeing a horse bled to death, and afterwards opening the body, he will be struck with the large amount of blood still remaining in the body, more especially in the large veins. The animal would seem to have died, not of a want of blood, but of a stagnation of blood; an empty heart, empty arteries—nothing to drive the blood out of the veins into the heart. A heart without blood is as useless as a pump without water. These questions are raised by the circumstance that there have been instances in which the injection of such a substance as milk, or salt and water, acted beneficially in restoring vitality. The qualities which are necessary in the injected fluid then may be simply a temperature about 100° Fahr., a neutral action on the tissues, the fluid to be injected of a specific gravity equal to the blood, and (a most important point) no tendency to clot the blood. The whole question is still open to investigation.

This concludes what I have to say about hemorrhage.

THE ELEMENTS OF SURGERY.

BY JOHN CHIENE, M.D., F.R.C S.E.

Surgeon to the Edinburgh Royal Infirmary, etc., etc.

LECTURE II.—(*Continued.*)

THE PROCESS OF REPAIR.—FIRST, IN A WOUNDED VESSEL; SECOND, IN THE INJURED TISSUES.

I have now spoken to you, gentlemen, of that which first demands your attention in any solution of continuity. I have considered it at this stage because I believe it to be the best introduction to the process of repair in the tissues after an injury. I now go on to speak of the *process of repair* as a whole. Suppose, in the case of any wound, you have at length succeeded in putting a stop to the bleeding ; you have sewed together the edges of the wound ; you have fixed on the dressing. What can you now do? Nothing. The remainder is a process of nature, and you can only here and there assist that process in the direction of healing. For two things may happen,—either the wound does not unite, the injury is followed by destruction and death of the part, or repair of the breach and recovery in the surrounding tissues is the result. We shall consider the former case in other connections ; our present business is with the latter.

The tissues are vascular, and every solution of continuity is associated with rupture of the vessels and consequent extravasation of blood. Extravasated blood has the power of coagulating. The rare exceptions to this do not interfere with the following argument : * *Repair of every solution of continuity takes place in every instance by means of blood clot.*

* Hunter describes cases in which the blood, after extravasation, remains in a fluid state. This may, perhaps, occur. My observations incline me to think that it is much rarer than is generally supposed. Undoubtedly many of the cases of so-called fluid blood can be explained on one of two suppositions : First, that the blood, having clotted, breaks down and liquefies. Second, that the clot *contracts* after forming, the serum tinged with colouring matter is squeezed out of the clot by this contraction ; that the serum is not reabsorbed, the clot floats free, and

This is simply a return to the teaching of John Hunter, with all the benefits of the knowledge which has been gained since his time. Hunter taught that fluid blood is a tissue "endowed with life." Life he defined as "the principle of self preservation." What he thought in regard to the function of the blood in the process of repair, may be learned from such quotations as the following :—" Extravasated blood is the bond of union." "The union of the broken parts to the intermediate extravasated blood," etc. "It is the blood and parts uniting which constitute union." "The blood so extravasated either forms vessels in itself, or vessels shoot out from the original surface in contact with it."

With these preliminary remarks, I shall now put what I have to say on this subject into the form of a series of propositions. These have been established in a separate connection, and may therefore for our purpose be called axioms. You know most of them already, but it is necessary to bring them before you in this connected way.

1. Where there is a breach in the continuity of a part, there is injury.

2. When a part is injured, there is a lowering of its vital powers.

3. When the blood is poured out, it is brought in contact with depressed tissue.

4. Coagulation of the blood is the result of contact with dead or dying matter.

5. Therefore extravasated blood coagulates.

6. The process of coagulation may be noted as consisting of three stages—(*a*) viscidity, (*b*) jellying, (*c*) contraction.

7. The colour of the blood clot is an accident. "The red colour of the blood is not an essential." (Hunter.)

it can not be felt by palpation. Fluctuation is felt, and it is said that the clot never formed. A clot floating in serum will sooner or later break down; it cannot organize. When a fluid is drawn off, in cases of extravastion, does it clot as normal recent blood does ? Hunter distinctly states that when blood does not clot after extravasation, inflammation is apt to ensue; in inflammation effusion is in excess of absorption. In such cases reabsorption of the serum from the contracting clot is not to be expected.

8. There is, therefore, besides the coloured, the colourless blood clot, as in the "buffy coat." (Hunter called the latter "coagulable lymph;" but people, in accepting the name, did not accept Hunter's explanation of the thing.)

9. The changes which take place in blood clot ending in organization, are interfered with and delayed by the presence of the coloured corpuscles.

10. The essentials in a blood clot are fibrin and white blood corpuscles.

11. To obtain these essentials, liquor sanguinis and white blood corpuscles are necessary.

12. In every irritated part you have effusion of the liquor sanguinis and migration of the white blood corpuscles. The elements of the colourless blood clot or lymph are present.

13. The nature of the changes which take place in a coloured clot will depend (*a*) on its surroundings, *i.e.* the tissue in which it lies ; (*b*) on its size ; (*c*) on its situation, *i.e.* whether it is exposed or not ; and (*d*) on its treatment.

14. If the tissues around are healthy, if the coloured clot is not of too large size, and if it is not irritated in any way, then it will organize, sooner or later, into connective tissue.

15. If it be poured out amongst unhealthy tissue, or if irritated by exposure or pressure, or if by its size it presses upon the neighbouring parts, becoming itself an irritant, then it will break down, liquefy, and be rejected or absorbed, or it may remain amongst the tissues, which are condensed by pressure and form a cyst wall inclosing the fluid, once blood.

16. The process of organization into connective tissue takes place in the following way :—(*a*) the blood clot becomes adherent to the surrounding parts ; (*b*) decolourization takes place ; (*c*) the cell elements in the clot increase by migration and proliferation : (*a*) the clot becomes vascular ; (*e*) formation of connective tissue takes place ; and (*f*) the process is completed by contraction. The process is the same as in the clot which closes the cut end of an artery.

17. This process of organization is accompanied by absorption of the fluid parts of the clot, of the red blood corpuscles,

and of any portion which, from the size of the clot, is beyond the reach of the living tissue from which the clot receives its vascular supply.

18. As a result of the preceding, blood clot results in either (*a*) recovery by organization ; or (*b*) death and liquefaction followed by ejection or absorption.

19. This recovery is equivalent to the formation of connective tissue, assisted by contraction and absorption.

20. Blood clot, in the process of organization, has always a tendency to become like the tissue near which it lies.

21. If the primary (coloured) clot does not form, or after having formed is removed, then the tissues are exposed and the migratory white blood corpuscles and the effused liquor sanguinis supply the element for the secondary (colourless) clot, coagulable or inflammatory lymph as it is called.

22. The changes which take place in the colourless clot are similar to those which occur in the coloured clot. The onward progress ending in organization is more rapid ; there are no coloured corpuscles to be removed. The colourless clot forms a thin layer on the surface of an open wound, or between the opposed surface of the wound, the walls of which have been brought in contact after all hemorrhage has ceased. In either case the tissues from which it receives its blood supply are in close proximity, vessels shoot rapidly into it—vascular colourless blood clot, when exposed, is termed granulation tissue, which covers the tissues and protects them from external irritation, and places them in the most favourable condition for recovery from the injury.

In the foregoing propositions is traced out the history of the general process. One or two varieties of wounds claim our special attention. For purposes of study we shall take comparatively simple cases, where we can approach to an exact enumeration of all the elements. In every large wound several varieties of healing may be noted.

Subcutaneous Wounds.—We have, first, the true subcutaneous wound, known as a contusion or bruise when occurring in the soft tissues, and a simple fracture when in the bone.

There is no solution of continuity in the epidermis, although the term "subcutaneous" is sometimes applied to wounds where there is a very small opening, as in the wound made in tenotomy. Good examples of a true subcutaneous wound are found in such common occurrences as a black eye and a kick on the leg. There is a tearing of the subcutaneous tissues, and blood is freely poured out. As a consequence, we have swelling and discolouration of the part. The irritation and depression have been followed by coagulation, and in connection with the coagulum the process of healing goes on. In some cases there is simple absorption, and this may take place very rapidly, as in the case of a black eye. In other cases, if the clot is not too large, if there be no irritation and if there be proper treatment, organization sets in as described above. But if the clot be too large, the central part will break down, become liquid, and be removed. Bone heals according to the same principles.

Open Wounds with Loss of Substance.—The second class comprises open wounds, accompanied by loss of substance. The wound is necessarily exposed for a longer or shorter time. Suppose it is for a very short time, so that putrefaction is prevented. There are blood vessels cut, and as it escapes the blood dries from exposure. In general the wound is dressed with a rag or protective of some kind, which, with the dried blood, forms a complete covering. Anything which will prevent the access of air, and act as a receptacle to the escaping and coagulating blood, will do ; and in the dog, an efficient protective is formed by a mixture of saliva, hair and blood. This drying produces a " scab." When the scab is removed at the end of a few days in a small wound, a new epithelial covering is found to have formed. The blood clot, protected from external irritation by the scab, commences to organize into epithelial tissue at the edges where it is placed next the epithelial tissue of the surrounding skin, and the process goes on until the whole surface of the wound is covered with a dry epithelial tissue, and the scab, not being retained in its place, falls off. So perfect is the adaptation of the new formation to the conditions of the organism as a whole, that in a case in which there was a blood clot with part of the mass beneath and part above the level of the original surface, and of the still existing surrounding surface, I have seen the new epithelial formation cut in like a knife through the clot, shaving off the redundant portion, which, in due time, fell off as a scab. This process is termed cicatrization, and the new epithelial tissue is termed a cicatrix. In the deeper parts of the clot, the changes ending in the formation of connective tissue are taking place. In order that the blood clot may organize under a scab, it is necessary that absorption should be in excess of effusion. If this is not the case, then tension will take place under the scab, the blood clot will break down, and suppuration will occur under the scab.

There may be cases in which the open wound is exposed for a considerable time, but yet by the adoption of antiseptic precautions (the consideration of which I shall take up fully in a few days), irritation by putrefaction is thoroughly prevented. Any case of operation with antiseptic precautions, in which the edges of the wound remain apart and the clot is exposed, may be taken as an example ; but I will refer to one instance in particular, which has a special interest for me as being the first in which I brought prominently forward the doctrine for which I am contending in these lectures. You will find the facts in the Lancet for July, 1875. They are, briefly stated, as follows :—I had resolved to make an experiment with a blood clot, to try its efficacy in the function of healing. A young man came to the clinical wards with a horn on his heel. I removed this, and the cavity which was left formed a kind of box, with the os calcis as the floor. I put on the protective over this cavity, took off the tourniquet, and the cavity was allowed to fill with blood. The blood clot filled the cavity, with the exception of one small hollow. The blood organized and epithelium proceeded to form over the surface. On the sixteenth day I scratched the surface of the new tissue ; it was vascular and bled, and we had the groove on the surface of the original clot filled up with a secondary clot. This second mass of blood passed through the same stages, and the ultimate result was a fair and sound cicatrix. It is interesting to observe that blood clot can organize on blood clot. This was the first case in which I observed this. We can, in consequence of this, make up for any insufficiency in the first clot; scratch it, after vascularization, and add to it.

If there is long exposure and no antiseptic precautions are taken, putrefaction will probably occur. Remember that I take putrefaction as merely an example of a local irritant. It is the most common one; its effect on the blood clot is due solely to its local irritant action. The blood clot in this case breaks down on the surface, and the putrefaction spreading deeper general disintegration commences. But in the deeper part of the clot, before the irritant effect reaches it, the normal process sets in. The putrefactive action going downward meets this developing tissue, in the stage of vascularization. Putrefaction can not go on in the presence of a healthy growing tissue, and therefore stops. The putrefactive products are thrown off, and the vascular lymph is exposed. The special name " granulation tissue" has been applied to this, but it is of less importance for you to know what it is called than to know that it is merely blood clot or " coagulable lymph," in the stage of vascularization. Though developed to a certain degree, yet it is still a very weak tissue, and great care must be taken of it. A slight irritation will break down the whole formation, and it is thrown off; the exudation of liquor sanguinis and white blood corpuscles takes place from the raw surface of the sore thus laid bare, lymph forms, and the process begins *de novo*. The lymph is colourless blood clot; the elements necessary for its formation are poured out from the tissue. These elements are liquor sanguinis, which supplies the fibrinogen, and white blood corpuscles, which supply the fibrinoplasten ; the result is the formation of fibrin, in the meshes of which white blood corpuscles are entangled.

We have now to study the changes which take place in granulation tissue. It consists of masses of cell elements resembling the white blood corpuscles. It is richly supplied with blood vessels, which communicate with the vessels of the part on which it lies ; it is surrounded by epithelial tissue, and it rests upon the connective tissue on which it has been formed. This layer of granulation tissue is seen on the surface of a healing ulcer or in an open wound. It may also be met with lining the cavity of a wound ; it does not necessarily communicate with the external air ; it is just blood clot or lymph which in the process of organization has become vascular, the deeper surface of which is in contact with tissue ; the superficial surface is unconnected with tissue. It is to be seen in any open wound which has not healed by scabbing, or by primary blood clot.

Two changes take place :—From the edge of sound epithelium a rapid development of epithelial cells goes on, projecting more and more into the centre of the sac, and among the round cells in the deeper parts of the clot conversion into connective tissue corpuscles, together with the formation of fibrous tissue from the intercellular substance, produce a subcutaneous layer over which the new epithelium forms. This process, you will observe, is similar in many respects to the process which goes on under a scab, and is also cicatrization. I should observe, at the same time, that during the conversion of a granulating surface into a cicatrix, contraction occurs, and the edges of the wound are drawn together. This fact of contraction, though frequently very unpleasant when the wound is on such a part as the face, can yet be turned to good account. If we have the raw surface of the wound not comparatively flat but disposed in a trench-like form, as in the case when a wedge of tissue has been removed we have the process of conversion beginning at the bottom of the cavity, and contraction taking place then brings together the two adjacent surfaces of the wound. This goes on, and the depth of the cavity constantly grows less, until you may come right up to the surface, the epithelium forming over all. This mode of healing is termed healing by "coäptation," and occurs in any trench-like wound, as in the operation for fistula in ano. There is always a certain amount of discharge from a granulating surface. It contains a number of very weak cells, and on account of their weakness and exposure there is as it were a high death-rate among them, and great numbers cannot maintain themselves at this low level of life. The discharge represents the death-rate. The granulation cells may be likened to individuals in a community :—an epidemic will increase the death-rate in a town ; improper treatment will increase the amount of discharge from a granulating wound.

Incised Wounds.—I have spoken of subcutaneous wounds, and of open wounds with loss of substance. I have yet to say something of a third class—open wounds without loss of substance, or incised wounds. Such happen in cases where there is a clean cut, and the knife passes into the tissues and is withdrawn without having carried away any portion. There are two modes of repair here, at one of which I have already hinted :

First—when you have the surfaces in contact. In every case of this kind the blood is poured forth from the ends of the cut vessels, and to a certain extent remains in the slit. The circumstances are all that we could desire for an experiment, and an incised wound of this kind, when opened after a certain length of time, shows the organization at one or other of its stages. This is one variety of union by the first intention. And I wish at this point to enter my protest against the assertion of the late Dr Macartney of Dublin, that what he means by union by the first intention is the same as what John Hunter meant. The error is the more serious, as it is countenanced by so high an authority as Paget. What Macartney means is simply this :—that if the cut surfaces are brought together immediately, all the parts which have been separated are at once brought into contact, more especially the cut mouths of the capillaries are united again ; and under these circumstances that, the circulation having received little check, time is only required for the soldering together of the broken parts—a process which is completed in a remarkably short time. Whatever may be the merits of this explanation (in my opinion it is an impossibility), it finds no support in the works of Hunter; and I refer you to his own works, and ask if a fair interpretation of his meaning will not lead you to the conclusion that by " union by the first intention " Hunter meant union by organization of blood clot.

Second. If you keep the surface of an incised wound apart, the colourless blood clot—the inflammatory lymph of Paget— covers the surfaces of the cavity. If the surfaces are then brought together, organization takes place, union takes place. If the surfaces are kept apart, the colourless clot becomes vascular, granulations form, and union takes place by coäptation and contraction. In the one case you have a wall of clot, with both sides of which the tissues are in contact, the epithelial formation forms the cope-stone. In the other you have the clot lining the sides of the cavity, lying on the tissues. The cavity heals from the bottom, the opposed surfaces come together by the contraction of the surrounding tissues, and coäptation of the surfaces takes place. The cavity decreases in depth until the surface is reached, and the epithelial formation joins the opposing edges of the skin— the wound is healed. Sometimes the epithelial formation may extend downward over the granulating surfaces, meeting the coäptation process in its upward progress. The result in such a case will be a depressed cicatrix.

I have gone rapidly over these different varieties of wounds ; it is not necessary to dwell more fully on them, because my object here is not to describe the phenomena, to explain the conditions well known to every surgeon. My belief is that much obscurity has arisen in consequence of a departure from John Hunter's teaching. The healing of wounds will be more easily understood by the student on the principles that I have tried to lay down. Healing by scabbing, union by first intention, union by second intention, are essentially the same process; the alterations are the result of the treatment of the clot, coloured or colourless, which has always a tendency

to form until the wound is healed. The natural history of
the clot will depend on its treatment and surroundings.

The healing process may be divided into two great divisions :
First.   *Healing by primary (the coloured) clot.*
Second.   *Healing by secondary (the colourless) clot.*

Under the first head comes scabbing; under the second
head comes healing by granulation, cicatrization and contraction, as in an ulcer ; healing by coäption, as in a trench-like
wound after a sinus is laid open. Union by first intention
may be the result either of the first or second division; of
the first, when the coloured clot is the uniting medium ; of the
second, if the wound is kept open for a few hours, until all
hemorrhage has ceased, until the secondary clot has formed—
the surfaces being then brought together, as Liston did
latterly in his amputation wounds. A granulating wound
may, if aseptic, be filled with blood by scratching the granulations, and the clot which forms may organize.

It is hardly necessary to say that in almost every large
wound, we may have examples of all these different varieties
of healing. Their difference is more apparent than real. If
I have been successful in my attempt to revive John Hunter's
teaching that blood is the uniting medium, that the changes
which take place in it result in a healed wound, then it follows
that *Repair of every solution of continuity takes place in every
instance by means of blood clot.*

THE AMERICAN PRACTITIONER.

AUGUST, 1879.

Certainly it is excellent discipline for an author to feel that he must say all that he has to say in the fewest possible words, or his reader is sure to skip them; and in the plainest possible words, or his reader will certainly misunderstand them. Generally, also, a downright fact may be told in a plain way; and we want downright facts at present more than anything else.—RUSKIN.

Original Communications.

THE ELEMENTS OF SURGERY.

BY JOHN CHIENE, M.D., F.R.C.S.E.

Surgeon to the Edinburgh Royal Infirmary, etc., etc.

LECTURE III.—INFLAMMATION.

INFLAMMATION—Causes: Exciting and Predisposing—Symptoms:—Local—Constitutional— "The Feverish State:" Three Varieties in Surgical Practice—Pain Fever—Waste Product Fever—Putrefactive Product Fever—Principles of Treatment—Rest—Prevention of Putrefaction—Removal of Foreign Bodies—Application of Cold—Counter-Irritation—Theory of Counter-Irritation Importance of Sick-Room Cookery.

Inflammation.—I have in the preceding lecture spoken of wounds and their repair, and I shall now proceed to the consideration of inflammation. It may possibly have struck some of you—especially those who have already attended a course of lectures on surgery—with surprise that I have hitherto not mentioned the word, and the more so that I may frequently have appeared to tread close upon the consideration of the thing. I certainly have intentionally refrained from using the word, and I certainly have spoken frequently of the thing as it is commonly understood ; but it is because I know that there are certain associations connected with the word that would

D

interfere with a clear apprehension of the thing. The word inflammation seems to imply something evil which it is as necessary to avoid as scarlet fever ; but when I spoke of the phenomena to be observed on the application of an irritant, you were prepared to give your whole attention to the process undisturbed by any association. Having done so you will not be led astray when you are told that the phenomena to be observed on the application of an irritant, as they have already been described to you, are the phenomena of the early stages of inflammation. My first care was to point out to you how clearly these phenomena were associated with the process of healing—how necessary they were to it ; and I should be inclined to confine the term inflammation to the phenomena when in excess, and when they appear as the symptoms—heat, pain, redness, swelling.

It is by no means an easy thing to restrict the meaning of any term which is in common use in a wider acceptation. Perhaps it is often undesirable to make the attempt. With regard, however, to the term inflammation, the word unfortunately is apt to be misunderstood by any one beginning the study of medicine. He has heard it always used to mean something evil—this is the meaning attached to it in common life. The surgeon, however, knows that without it no wound would ever heal, and he welcomes it in his early stages. He desires, however, that it should be kept within bounds; he knows that if it goes too far, instead of being the means by which a wound heals it prevents that healing. No wonder that the junior student is bewildered. An attempt has been made in these lectures to avoid a misunderstanding by speaking of the process of repair before entering on inflammation. I now proceed to speak of the causes, symptoms, and treatment of inflammation.

Causes.—The causes which give rise to inflammation are divided into exciting and predisposing. The division is not a scientific one, but it is useful. An example will bring out what is meant by it. Two men receive the same injury, and are placed under the same treatment: the one recovers, and the other dies. The circumstances of the two men from the time of receiving the injury were the same, and the difference in the result must lie in the state of the two men before the injury. The predisposing causes led to the difference in the effect. Thus the whole effect and the ultimate result will be determined by the nature and extent of the injury which is the product of the exciting cause, and by the circumstances and constitution of the patient, at the time he received the injury—the predisposing cause.

Exciting Causes.—(*a*) Mechanical injury. This requires no explanation. (*b*) Chemical. An acid which acts on the tissues, such as nitric or sulphuric acid, may serve as an example. (*c*) Heat or cold. Burns and scalds are the effect of the first ; frost-bites and chilblains of the second. (*d*) Putrefaction. This is one of the most fertile causes, and with it is closely associated the question of antiseptic surgery. (*e*) Foreign bodies. You must be prepared to give a wide interpretation to this phrase. Pus, which collects in the tissues, is a foreign body as much as a leaden bullet. Indeed, it is often not so much what the foreign body is in itself as what it carries with it—the germs of putrefaction—that constitutes the exciting cause.

The *predisposing causes* are—(1) Constitutional peculiarities. Such are, (*a*) hereditary tendencies—*e.g.* to gout or scrofula; (*b*) bad habits, such as intemperance; (*c*) starvation; (*d*) proneness to excited action or over-sensitiveness : one man will keep on a mustard poultice for half an hour, while another can not bear it for more than five minutes; (*e*) altered or weakened nerve power, as when bed-sores form in cases of paralysis; (*f*) certain states of the blood, as when carbuncles form. Notice that this last is a refuge for our ignorance, and may mean almost anything. (2) Previous attacks. When you once have had a part attacked, there will always be weakness there. When once an inroad has been made by such diseases as gout and rheumatism, it is very easy for them to return.

Symptoms.—The symptoms I am about to speak of you will not find so well marked in all cases as I shall describe to you.

I am speaking of typical cases, and from the standard thus
set up there are numberless variations. One patient has one
symptom well-marked, another has another; in one part of
the body you will find inflammation characterised by the
development of one symptom, and in another part another
symptom attracts your attention. The cause of the inflammation, the constitution of our patients, and the seat of the
disease, are the factors which determine the relative development of the symptoms in any one case.

The symptoms of the typical inflammation are either (first)
local, or (second) constitutional. Each class deserves careful
consideration, as about each there is yet a great deal of which
we know nothing.

Local.—These are generally summed up in the old formula
—redness, heat, swelling, pain. Some of these terms do not
precisely describe the actual phenomena, and I prefer to
present them in a modified form.

(*a*) *Alteration in colour.*—The colour toward which an
inflamed part tends is red. Be the shade of colour what it
may, the alteration is always owing to the greater quantity of
blood in the tissue. It is congested; and the amount of
congestion will depend on the vascularity of the part, the
strength of the patient, and the acuteness of the inflammation.
On these conditions, therefore, will depend also variations in
the intensity of the colour: when there is biliary derangement,
the colour of the inflamed part tends towards yellow.

(*b*) *Swelling.*—This will be more prominent in those tissues
in which there is the greater room for it. For example, in the
conjunctiva the swelling is great, and so also in the testicle;
but in such hard dense structures as bone, cartilage, the cornea
and fibrous tissue, there is very little swelling. The special
cause of this inflammatory swelling has been alluded to
already more than once. In an irritated part there is effusion
of the liquor sanguinis, together with migration of the white
corpuscles of the blood; and this infiltration in the tissues,
together with the subsequent proliferation of the cellular
elements, causes the swelling.

(*c*) *Heat.*—You must understand by this not the mere feeling of increased heat in the part, but the actual fact that there is greater heat developed in it. We owe the demonstration of this to Simon and Montgomerie. It was ascertained by them that the temperature in the blood of an artery going to an inflamed part was less than that in a vein leading away from the part, and that the temperature in the latter was less than that in the seat of inflammation itself. The following illustration will show you the meaning of this. The air in this room while that fire is burning takes a course towards the fire-place and up the chimney, and you would find the temperature of the room, the chimney and the fire itself, roughly speaking, in the same relation as those of the artery, the vein, and the inflamed part. The explanation of the relation in this latter instance is the same as the explanation in the former. The inflamed part is the seat of the production of heat. In the changes that are going on there—changes tending to degeneration—there is continual oxidation to a much greater extent than is usual in the normal tissue, whence the increased heat.

(*d*) *Altered sensibility.*—The word "pain" does not express all that is meant by this. When you have inflammation of the ear, you have tingling ; and when the eye is inflamed, flashes of light pass across it without an outward cause. But the organs of special sense being put out of question, pain is the general form of manifestation of this symptom. It is of various kinds, which can not be described otherwise than by the graphic epithets which are used by those who have felt them — starting, throbbing, aching, burning, gnawing, etc. Very frequently the intensity of the pain varies with the denseness of the structure affected. The pain in acutely inflamed bone or tendon is excessively severe, while in cases of inflammation of the loose cellular tissue it is comparatively mild. This altered sensibility in all its modes is due, no doubt, to the pressure upon the sensory nerves. The pressure on the sensory nerves will be most acute in the dense tissues, where the effusion is so strictly confined ; while in the looser tissues, where the added constituents have room to expand, the pressure is not felt so acutely. While the sensory nerves are affected in this manner, it can not but be that the vaso-motor nerves are also influenced, and the peculiarities in the vascularization of an inflamed part will be partly due to this latter cause.

One form of pain I must not omit to mention here. When there is disease of the hip joint, there is pain experienced in the knee. When there is inflammation of the bladder, there is pain at the point of the penis. These are examples of what is called sympathetic pain, and are due, no doubt, to a peculiar arrangement of the nervous mechanism. Fuller dis-cussion of the subject I must defer till a later period, when I speak of counter irritation.

(e) *Modification in function.*—This is a result of the inter-ference with the nutrition of the part. An inflamed muscle can not contract: an inflamed bladder can not hold water; an inflamed eye can not see, and so on through all the organs of the body.

Constitutional symptoms.—It is especially about this class of symptoms that our ignorance is most manifest to ourselves. But if any of you desire to attempt to let in a little light upon the obscurity that reigns here, I will say in passing that I know of no other way by which best to approach this study than a sound knowledge of physiology. I have had occasion to magnify this study to you before, but in the consideration of these constitutional symptoms of inflammation, especially you will not advance a single step in explanation unless forti-fied by a careful discipline in physiology.

The constitutional symptoms are summed up generally in the expression, *the feverish state.* There are great variations in the character and intensity of this state, but we may with confidence refer them to the interactions of the three condi-tions mentioned a little ago—the cause of the inflammation, the constitution of the patient, and the seat of the inflamma-tion. We have a specific poison, and we have the constitu-tion on which the poison acts; and a fever, like every other product of nature, will undergo modifications both from the seed whence it sprung and from the soil where it was reared.

The feverish state is the collective name for a number of familiar symptoms. I shall take these *seriatim*.

First, there is a rise in temperature, accompanied by a rigor and increased thirst, with want of appetite. Secondly, the secretions are scanty, *i.e.* the skin does not act, but becomes hot and dry, and the normal amount of perspiration is absent; the bowels are constipated on account of the lessened secretions from the lubricating glands; the urine is small in quantity and high-coloured, from defective action of the kidneys; the tongue becomes furred and unclean, because the secretion of the salivary glands is affected. All these are manifestations of the second symptom. Thirdly, respiration and the heart's action are increased. The latter of these phenomena is, as we shall see, the result of the former. Fourthly, certain nervous complications take place. Languor and severe headache set in, and are followed by a confusion of ideas rising maybe to a state of delirium.

This condition may end in death as a result of exhaustion, or the patient may be lifted up again, and recovery take place. The recovery is preceded by a critical evacuation. The secretory organs begin to act vigorously. A profuse perspiration breaks out; the bowels are freely moved; even unwonted passages may be found for the renewed life of the body, and there may be bleeding from the nose.

Causes of the constitutional symptoms.—To assist in seeking for a proximate cause of all these phenomena, we have a few established facts in physiology and pathology. There is a process of increased oxidation going on in the inflamed part, and associated with this an increased production of heat and an increased production of effete matters; there is an accumulation of these waste products in the blood and deterioration thereby, and this is furthered by the defective action of the secreting organs described above; and lastly, there is nature's attempt to throw off the mischief. Let us see how far these conditions will lead us to explain the proximate causes of the symptoms mentioned above.

(1) *Rigor and rise of temperature.*—We have all suffered from this, and know well what it is. When you have a rigor, there is contraction of the vessels at the periphery of the body, and dilatation of the vessels in the internal viscera. This affects the mechanism which regulates the addition and withdrawal of heat in the body. The liver is the great heat producer, assisted to some small extent by the muscles; the skin is the great heat withdrawer—of every one hundred degrees of heat withdrawn seventy-seven degrees being by way of the skin. In the above circumstances, then, there is contraction in the seat of common sensation, whence the feeling of cold and shivering; but there is dilatation and increased activity in the heat-producing area, whence the rise in temperature. You may compare the body to a room. You may increase the temperature of the room in one of two ways, either by a larger fire, or by preventing the heat that is formed, from getting exit. Both of these causes may be at work at the same time. Increase the size of the fire, close the windows. The liver is the fire, in it you have increased vascular activity; the skin is the closed windows, in it you have vascular contraction. But why there should be contraction peripherally and dilatation in the viscera, we do not know. We can understand why one should feel cold after dinner, because we are aware of a definite cause for visceral activity; but we are aware of none in a case of blood poisoning.

(2) *Scantiness of the secretions.*—The blood is heated and deteriorated, and the skin, kidneys, etc., do not receive a proper supply of good blood, and their normal function is impaired. The passage of heat also is affected by this diminution in the excreta. After a time there will be a poisoning of the nervous centres, and the nervous action on the secretory organs will still further complicate the process.

(3) *Increase of respiration and heart's action.*—This is simply nature's attempt to throw off the mischief. The lungs in health throw off about twenty per cent. of the heat that is withdrawn from the body; they now attempt to do more, and to take the place of the skin. The respiratory and

cardiac ganglia work longer than the other, and are able for
their work when the others have failed, and therefore they
try to do the work of the others also.

(4) *Certain nervous complications.*—This simply means an
advance of the blood in its deteriorated state to the nervous
centres and a poisoning of them.

Such, so far as we can ascertain, is a probable explanation
of *how* these symptoms come about; it remains that we consider the "*why*" of the state as a whole. What is the cause
of the feverish state?

Causes of the feverish state.—There are three causes present
of which we know something. It can not be said that these
three causes are present in all instances. Indeed, we actually
count three distinct types of fever, each marked by the
predominance of one of these causes, and the comparative
inactivity of the other two. These are the *pain fever*, the
waste product fever, and the *putrefactive product fever*.
Though they are essentially quite distinct, yet in the majority
of cases they are more or less coëxistent.

(*a*) *Pain fever.*—You may call this, if you like, tension
fever, as it is tension which is the objective fact when pain is
produced. It is of very common occurrence, quite apart from
the rest. Suppose that one of you suffer from whitlow—an
inflammation of the flexor tendon on the front of the finger.
When the pain grows intense, you are seized with a rigor,
and a rise in temperature occurs, which continues until the
surgeon cuts down upon the seat of inflammation, and affords
local relief. The seat of inflammation is under the fascial
sheath, and as you know there is very little space under this
unyielding material to give the effusion room to spread.
Great pressure is thus caused upon the sensory nerves. This
stimulation is communicated to the centre in the medulla
oblongata, from which impulses sent along the vaso-motor
nerves induce the rigor and consequent rise in temperature.

(*b*) *Waste product fever (wound fever or traumatic fever).*—
In this kind of fever you have a distinct poison passing into
the blood; the manufacture takes place at the seat of the

local injury. There is increased oxidation, and the effete products of this process being abnormal, have no immediate normal outlet. They remain in the blood and tissues, and are conveyed to other parts of the body chiefly by means of the lymphatics. The constitutional disturbance is met with from the second to the eighth day after receiving the wound. During this time the surface of the wound is an absorbing surface, and the mouths of the lymphatics are open. After this period generally granulation commences, and with the commencement of this the wound ceases to be an absorbing surface; the mouths of the lymphatics are closed, and the surface begins to cast off.

This posion is not a strong one, and that the effects may be appreciable a large dose is required, and therefore a large surface of wound. The effects lasts no longer than the period of production of the effete matter. This condition disappears with the appearance of granulation; absorption only takes place when the wound is recent, when the lymphatics and blood-vessels are open; when they are closed, then the wound fever ceases. Such is the waste product of traumatic fever.

(*c*) *Putrefactive product fever.*—This is by far the most important, because it is the most dangerous, kind of fever which the surgeon has to deal with. It may occur at any time until the wound is healed. Unlike the traumatic fever it does not require a large dose of the poison, nor on that account a large absorbing surface. It is a powerful poison. Besides the amount of the dose, the constitution of the patient is an important factor in the process. The poison enters either by the lymphatics or by the blood vessels. Two types of putrefactive poisoning are met with. The first is septicemia; in this the course of the fever is very rapid. In the second (pyemia) it is slow, and a patient may linger on for weeks under an attack of this variety. In one who has been strong and healthy up to the time of receiving a wound, and who is inoculated with the putrefactive product, the fever is apt to take the acute form, while a weakly patient wastes away with a fever of the low or chronic type. In all probability the septicemic variety enters by the lymphatics; the pyemic variety enters by the blood vessels. Septicemia occurs soon after the infliction of the wound; pyemia at a later period.

The treatment of pyemia or septicemia is, I am sorry to say, unsatisfactory; it is a most dangerous disease, and the majority of those who have died of the effects of operations have been killed from this cause. But if the treatment is a subject for despondency, on the other hand the prophylaxis of the disease may be so conducted as to give the very best results; and this will form a topic for fuller consideration at a subsequent stage in these lectures. I have never known a patient suffer from pyemia or septicemia in whom the wound was aseptic.

Principles of treatment.—In treating inflammation, as surgeons, remember that you have the advantage of knowing that at a certain time, viz., after a wound has been received, and in a certain place of the body, viz., the part which has been wounded, inflammation may with some probability be looked for. Preventive measures, then, ought always to have a large share of the surgeon's attention in the treatment of inflammation. And as in most cases the inflammation is kept up by the continuing presence of the original irritant, the principles of curative treatment will be much the same as those of preventive treatment. In the following observations you will understand what I say to apply in great part to the one as well as to the other.

Rest.—The first great principle which we must learn to apply is that of rest. Your study of physiology must be brought into play to teach you what is rest and what is unrest for any particular organ; for to rest an organ simply means to prevent it from performing its function either by rendering the performance unnecessary altogether, or by calling into action other agencies which may take the place of the one in question. Rest may be divided into general rest and local rest.

One of the great difficulties you will meet with in practice is in dealing with patients of an over-sensitive disposition.

Your first aim must be to reduce these people to such a condition of rest as will give the affection they are suffering from
a fair chance of recovery; and, for this purpose, no medicine
is so valuable as opium.   The caution which you most need
in the administration of this drug is that you must beware of
giving too small doses.   Small doses stimulate, and it requires
a large dose to obtain the sedative action.

The means you employ to secure local rest may be either
mechanical or physiological.   When you apply a splint to a
fractured limb, in order to overcome and prevent the action
of the muscles, the means are mechanical ; when you exhibit
purgatives and diaphoretics, in order to relieve the kidneys
of their work, the means are physiological; when you pack
the penis and testicles in cotton-wood, and at the same time
administer bromide of potassium or camphor, the means are
both mechanical and physiological.   There is no definite line
of demarcation between these two methods.   The body is a
living organism, and the performance of every one of its functions is a physiological process; but at a number of points
we can observe the application of mechanical principles, as in
the motion of a bone pulled upon by a muscle, and conse-
quently when we wish to interfere with an action of the latter
kind, we can, to a certain extent, adopt mechanical means.

In many cases where the performance of the function is not
an essential part of the vital process, and where it is under
the control of the will, you will be assisted by nature in your
efforts.   In an inflamed joint there is acute pain felt in the
act of moving, and on that account the patient will volun-
tarily exercise all his care to keep it at rest.   Pain is often an
excellent splint.   But when the control of the will over the
muscles is diminished, as in sleep, this aid is gone; hence
the intense pain felt in an inflamed joint when the patient is
dropping off to sleep.   Of course sleep must be secured
at all risks, and the bandages, weights and splints we apply
have reference in such cases mainly to the state of the patient
when asleep.

The amount of energy in the body is definitely limited,

and sometimes in order to relieve one organ from tension and work, it is sufficient to put an extra amount of pressure on another. The typical example of this in cases of over-worked brain. The remedy is active bodily exercise. When engaged in playing football, as you know, it is actually impossible for you to think in a sustained and consecutive fashion. Your brain, therefore, "gets a rest," when your limbs are active, or when your bodily senses are occupied as your eyes are with a change of scene, your thinking brain gets a rest.

The second principle I would recommend for your attention is the *prevention and cure of putrefaction;* entering into this subject involves the detailed study of antiseptic surgery, which I will defer until I come to suppuration.

Thirdly, we must take steps for the *removal of foreign bodies,* which do harm by causing unrest. These I would divide into two classes. There are the true foreign bodies, those which enter from without. Such are pieces of cloth, bullets, splinters of wood, etc. These chiefly derive their power of doing mischief, and therefore their importance, from the circumstance that they generally carry along with them the seeds of putrefaction. An aseptic bullet may get into a man's body, and lie there for years without doing any harm. But besides these true foreign bodies, there are others which it would not strike you to call foreign bodies. Pus is a foreign body; the effusions in hydrocele or synovitis are foreign bodies. They have no business where they are, and they are performing no function. They are merely pressing on the neighbouring tissues and acting as irritants. These foreign bodies are produced from the tissues in which they lie; but that is not the principal fact to be noted regarding them. *They are constantly being added to;* and their action as mechanical irritants is therefore ever growing more severe.

These obstacles to recovery and excitants of inflammation must be removed; and the removal of the third class I have mentioned requires particular methods. Take a case of synovitis where the knee-joint is affected. The treatment often is to take a fine trocar, and opening with this into the joint let out the contained fluid. It may probably be necessary to do this more than once. The remedy is intermittent in its action, in its power for good. If it is necessary to give the joint continuous rest for some time, then a drain is inserted to draw off the fluid continually. This is an example of the curative action of a drain. You have an example of its preventive action when it is inserted into an operation wound before sewing it up. Do not suppose that it is always necessary to drain a wound externally. There are great surfaces of the body which in a healthy condition are absorbing surfaces, and it is sometimes possible to take advantage of their existence to drain our wound. For example, a wound near the knee-joint may be drained into the knee-joint and completely stitched up. The wound in ovariotomy often requires no external drain, if the peritoneum is healthy. Might not the wound made in an operation for hernia be drained into the peritoneal cavity? In a case of femoral hernia in the female, it is very difficult to keep the wound aseptic, since it is so near the external opening of the urethra. If, then, we could drain our wound into the peritoneal cavity, we might have healing of the external opening in forty-eight hours by the first intention.

Fourthly, *the application of cold* will sometimes be found of use. When inflammation has reached a marked degree, cold may injure, and generally it must only be used at the very outset. Again, it is only useful in superficial inflammation, except in the case of inflammation within the cranial cavity, where it is used with excellent effect. The amount of cold necessary may be set up either directly as by ice, or indirectly by evaporants.

Counter-irritation.—We know fairly well the modes of action of the foregoing therapeutical agents; but I have to deal now with one very important principle of treatment of which we know very little. I shall present to you one or two facts which may serve to reveal the meaning of *counter-irritation.* Inflammation of the lungs is relieved by the application of a blister to the chest. An inflammation in the knee-joint is treated by a blister to the skin, and a case of ulcerated cartilage may also be treated by inducing an ulcer in the superficial tissues.

To speak shortly of the explanation of these facts, we may lay it down as a safe generalization that two inflammations of a different organ or tissue cannot exist comfortably in the body at the same time. To take an example: A person is suffering from mumps—an inflammation of the parotid gland. Suppose the person a male. One morning when he wakens he finds the inflammation in the gland gone, but his testicle is swollen. Suppose the person is a female. In her case the inflammation disappears coincidentally with the appearance of ovarian irritation. An individual is suffering from erysipelas, and while there is great inflammation of the scalp and face, the contents of the cranium are evidently untouched, and the patient's mental condition is quite sound. In a day or two you find that the erysipelas has abated, but that there is a depressed nervous condition and confusion of ideas—the brain is evidently affected. All such occurrences are referred to under the name metastasis. Hunter treats of much the same thing, but calls it sympathy.

Now surgeons take advantage of this law of metastasis when they apply a counter-irritant. Recollect what I told you of the action of an indirect irritant. It acts through the nervous system; and this hint gives us our first position when we try to frame a theory of counter-irritation.

(1) The mechanism connecting the two parts concerned must be a nervous one, and you will remember what I said about such a case as pneumonia, where the intervention of the pleural cavity between the parts affected forbid the supposition that anything else than an indirect connection through the nervous system is the basis on which the inflammation is caused at the first, and I now add, is cured afterwards.

(2) This connection is through the cerebro-spinal system.

(3) What is termed a vaso-motor mechanism, which establishes the connection between the sensory and vaso-motor nerves, is the special bond of union.

(4) There is an intimate connection through this system between the superficial and the deep parts of the body; pain in the shoulder, a consequence of disease of the liver; pain under the mamma in ovarian irritation; supra-orbital neuralgia from a carious tooth.

(5) The amount of nerve force in the body and the amount of blood supply in the body, at any given time, are fixed quantities. In other words, if an extra amount of nerve force or of blood supply is expended at one part of the body, it must be taken from some other part.

(6) When you have inflammation in any part, there is an alteration in the nerve force and in the blood supply expended at that part. This alteration consists in addition at the inflamed part, and as a consequence of this there must be substraction from some other part.

I have now to bring these facts to bear upon the explanation of the action of a counter-irritant.

The explanation of counter-irritation which I have to offer is not held by me as proved. It is a scientific hypothesis put forward as a probable account of what takes place, and at any rate as furnishing ground to go upon in further research. These theories, which are put forward in the progress of science, have this value independent of their positive truth, that they point out what experiments should next be made for the verification or refutation of them; nor does the refutation of them prove them valueless, for it may turn out to be a most important step in the approach to the real truth. It will be your special work, gentlemen, to test the theories of those who have preceded you; and no more suitable legacy can you leave to those who will follow you than a carefully thought-out and clearly-stated hypothesis.

The germ of the following speculation is to be found in a paper by Mr Hart in the London Practitioner, November 1878, on "Sympathetic Pain." It is applicable to those cases in which we have no direct vascular communication between the part inflamed and that part to which the counter-irritant is applied. As I hinted before, counter-irritation may be looked upon as a mode of bleeding. Now, in such a case as that of the kidney, where we have direct vascular communication between the affected part and the part to which the irritant is applied, the bleeding is direct, so to speak. By producing hyperemia in the skin and parts immediately beneath, we withdraw blood from the kidney. But in such a case as the pneumonia, where no direct vascular connection exists with the breast, the bleeding is produced by the intervention of the nervous system, and making use of the fact that hyperemia of the nervous centre is associated with hyperemia of the inflamed part. So may be explained the connection between the mamma and the ovary, which we know, from their forming two of the organs of reproduction, must be closely associated in the nervous mechanism. So also may be explained why gonorrhea may be cured by giving rise to orchitis. You will understand what the surgeon does if I take this latter instance, and say that we may cure gonorrhea therefore by giving rise to orchitis.

Take one or two more examples. Not an uncommon result of gonorrhea is a bubo or inflammatory enlargement of one of the glands in the groin. In the early stages of the affection a common method of treatment is by a blister. This has one of two effects : it either causes suppuration or absorption. Which of these effects is brought about depends on the way in which the blister acts. It may act directly on the inflammatory swelling, and having itself an irritating effect, will simply hasten the process; this brings on suppuration. Or it may act on the neighbouring skin and cause depletion in the inflammatory area, thus acting as a counter-irritant; in this case it will produce absorption. Some time ago I tried the following experiment: A man came under my charge with a bubo on each side. I applied a blister to each, putting it directly over the one, but in the case of the other adjusting it in a horseshoe from above the bubo so as to catch the sensory nerves that came from the gland. The first suppurated, and the second was absorbed. Again, consider how a

F

dose of castor-oil acts when it is administered for inflammation. The active principle of the medicine is an irritant, and
produces its effect on the mucous membrane of the stomach
and intestines. These, on examination in the body of an
animal to which castor-oil has been administered before death,
are found to be congested, and doubtless this acts in depleting
the congested tissue elsewhere in the body.

The end we have in view in counter-irritation is the relief
of vascular tension, and however it may be that it is so, yet
no doubt we do secure this end just as surely as if we were to
bleed from the median cephalic vein. Yet we secure another
end—the restoration of the secretions. By removing the
hyperemia, we restore the tone of the affected organ. This
when we deal with a particular organ. How it is that on the
cessation of a fever all the secretive functions recover their
tone at once, and how it is that by the exhibition of such
remedies as castor-oil or Dover's powder this result follows,
we do not know.

It is now generally allowed that a counter-irritant acts
through the nervous system. The nervous mechanism is a
combination of an efferent impulse along the sensory nerves
of the part to which the counter-irritant is applied. The sensory nerves are in connection with a vaso-motor centre; as a
result of the afferent stimulus a change takes place in the
vaso-motor centre; the result of this change is an alteration
in the conditions of the walls of the blood-
vessels which are under the command of
the vaso-motor centre.

Let me illustrate this by a rough drawing:
A mustard-blister is applied to the skin at S;
an impulse passes along the sensory nerve
(S N); a change takes place at vaso-motor
centre (V M C); the result is an efferent
impulse along vaso-motor nerve (V M N);
the result is a change in the size of the
blood-vessel in the skin (B). Dilatation
takes place.

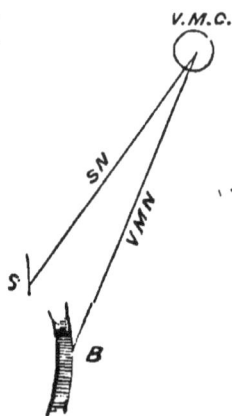

Let us now suppose that we have an inflammation of the kidney. The blood-vessels of the kidney are dilated with a slow flow of blood through them. Let the word *congestion* be used for this condition, to distinguish it from dilatation of the blood-vessels with a quickened blood flow. For this condition let us use the word *determination;* in both of these conditions there is dilatation. In the first, congestion, the function of the kidney is improperly performed, the kidney is in a state of inflammation. In the second determination, the function of the kidney is increased, when the blood-vessels of the kidney recover their tone there is a condition which, relatively to the state of determination or congestion, is one which we speak of as anemia. The following little diagram may assist you :

The kidney (K) is in a state of inflammation. The blood-vessels (B') of the kidney are congested. A mustard-plaster is applied over the skin (S) of the loins; the result is a change in the vaso-motor centre (V M C), which rules the skin vessels (B). These blood-vessels dilate. We have also, as a result of the counter-irritant to the skin, a diminution in the size of the kidney blood-vessels (B').

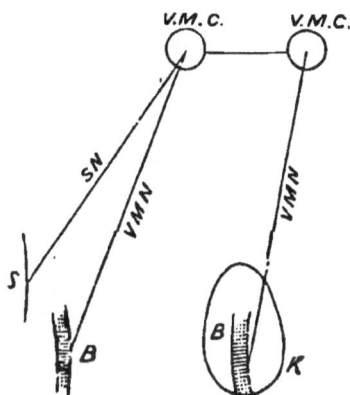

From a state of congestion they pass into a state of determination, with a free flow of urine, and from that to a normal condition, which is relatively one of anemia. In consequence of the mustard, a change has taken place in the vaso-motor kidney centre (V M C'), which rules the vessels of the kidney.

What is the nature of this change ? The vaso-motor centre is an organ. An organ as regards its functions may be in one of three conditions : it may be, first, at rest ; second, its function may be increased; third, its function may be abrogated. Let us consider the condition of a muscle when in these different conditions functionally. When a muscle is inflamed, it is congested ; when active, as in tetanus, there is determination: when at rest, there is a condition which relatively is one of anemia. Take another organ, the brain : when the brain is inflamed, its vessels are congested; when active, as in thought, there is determination; when at rest, as in sleep, it is anemic. These three conditions are associated with and probably due to an alteration in the blood supply of an organ, such as a muscle or the brain. If true of the whole brain, may it not also in all probability be true of different parts of the brain— may it not be true of a vaso-motor centre?

Let us now apply these conditions to the poultice in the inflamed kidney. When the kidney is inflamed the vessels are dilated, the vaso-motor kidney centre hast lost command ; it, like the inflamed brain or inflamed muscle, is congested. Apply your poultice; it acts on the skin centre; it is altered as evidenced by the condition of the skin vessels ; it is congested. Where does the blood come from ? From the neighbouring parts — from the vaso - motor kidney centre. Vascular tension is relieved; the centre regains its power; the kidney vessels are restored to their normal condition.

This explanation appears to me to be founded on a rational basis. I do not know the meaning of any such term as a change in tone of a vaso-motor centre; its tone or function must depend on its anatomical condition. When we remember the transient effect of a poultice, requiring its repetition or continuance, it must be evident that the anatomical change in the centre is temporary in its nature. It must be the blood supply which is at fault, either of improper quantity or of bad quality. No doubt the latter cause is at work in the feverish state, but the condition is a general one. The effect of the counter-irritant is local in the case of the kidney at any rate; it can not act by directly improving the quality of the blood; it acts by changing the quantity. Remember how minute the organ is: a single minim of blood plasma drained from the vaso-motor centre will have an effect on that organ equivalent to the withdrawal of a couple of ounces from the biceps muscle.

Let me try and show by another drawing the nature of the change in the vaso-motor kidney centre, which, when congested, is bled into the vaso-motor skin centre.

There is a microscopic bleeding at A in direction of arrow, from kidney centre to skin centre. In the case of the kidney, the counter-irritant relieves the kidney in another way. You can inject the skin of the loins from the renal artery; there is a free anastomosis between the kidney and the skin. The skin vessels are dilated; the blood passes from the kidney B' to B the skin. There is, therefore, a microscopic bleeding from B' to B, relieving the vascular tension directly.

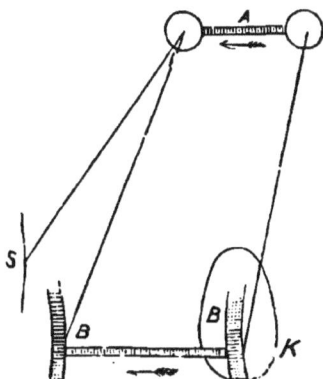

It may be asked, Is it necessary to theorise about the microscopic bleeding at A?—Will the microscopic bleeding at B not be sufficient to account for the good effect of the counter-irritant! Consider the lung when it is inflamed—the condition is relieved by poultices to the chest-wall. In this case there can be no direct microscopic bleeding; there is no direct anastomosis between the vessels of the skin and the vessels of the lung. The pleural cavity intervenes. Indirectly through the whole mass of the blood, the lung may be relieved. This can not be the main factor. In this case the good result must be due to a change in the vaso-motor mechanism. The bleeding must be *microscopic*, the lung centre being bled to supply the skin centre.

A counter-irritant does good by relieving vascular tension. It is a variety of blood-letting. It acts in two ways:—first, a microscopic bleeding from vaso-motor centre to vaso-motor centre; second, a microscopic bleeding direct in some cases, as in the kidney through the lumbar anastomoses, indirect in others as in the lung, where there is no direct anastomosis.

A dose of castor-oil acts as a counter-irritant. When castor-oil is administered, the intestinal mucous membrane is congested—the oil is directly or indirectly an irritant. The blood is drawn from other organs to supply the intestinal vessels. A purgative, then, may be considered a counter irritant; so also a diaphoretic or a diuretic : hence their great value in inflammation.

It is hardly necessary to say that by restoring the arrested secretions, they remove they waste products which have accumulated in the blood, and by removing the serum of the blood they still further relieve the general vascular tension.

Metastasis gave surgeons the hint which directed their attention to counter-irritation, and its value in practice. No doubt counter - irritation is an artificial metastasis. The parotid is inflamed; its vaso-motor centre is congested. This congestion spreads to neighbouring centres. The centre of the testicle is implicated; it becomes congested, loses command of the vessels of the testicle, and the testicle inflames. The parotid centre is bled—the parotid inflammation is relieved.

It is hardly necessary to do more than mention other means which may be adopted in relieving and curing inflammation. I have tried to show that a counter-irritant acts by relieving vascular tension, and has, to a great extent, replaced the different methods of blood-letting by leeching, cupping, scarifying, and general blood-letting. I have also pointed out that diaphoretics, diuretics, or purgatives, act as general counter-irritants; they also relieve vascular tension. They also act by restoring the functions of the secretory organs—the skin, the kidney, and the bowel. When the pain is great, sedatives are of value—a local sedative as an opium fomentation, a general sedative as opium, hyoscyamus (of special value in inflammatory affections of the genito-urinary system), chloral and bromide of potassium. When the patient is asthenic, stimulants are indicated.

We must not forget the great value of proper diet, or as it is called "kitchen physic." So important do I consider this, that your attention will be specially directed to diet in the lectures on sick-room cookery, which will be delivered by an expert on this subject, in which you are taught how to make the different articles which are most suited for the sick-room. Every physician should be able to cook all the simples needed in a sick-room. Such a knowledge is useful; it gains the confidence of the patient and his friends. Ignorance of these things often brings the medical man into discredit. We are only able to judge of our neighbour's knowledge by our own standard of knowledge. Ignorance will not lower you in the estimation of the ignorant man. Whatever a man knows, he thinks should be generally known. The mother of the patient will not interfere with your medical theories or your medical practice, however wrong they may be; she will see that her child swallows whatever you order; she is ignorant of these things, and your ignorance (if present) will be undiscovered, but it is altogether a different affair if you order her to give the child some food, and are unable to give her the necessary instructions for its preparation. She may feign ignorance to test your knowledge. If she finds you ignorant in one thing, then she is apt to argue that you are ignorant in all. Study sick-room cookery.

The next lecture will be devoted to suppuration and the principles of antiseptic surgery. I have a reason in associating suppuration and antiseptic surgery. It is a common mistake to suppose that when suppuration occurs in a wound, the antiseptic management has failed; it is forgotten that putrefaction, which is prevented by antiseptic precautions, is only one cause of suppuration. The use, for instance, of a powerful antiseptic will cause suppuration; tension is another cause; a severe inflammation a third cause.

THE AMERICAN PRACTITIONER.

NOVEMBER, 1879.

Certainly it is excellent discipline for an author to feel that he must say all that he has to say in the fewest possible words, or his reader is sure to skip them; and in the plainest possible words, or his reader will certainly misunderstand them. Generally also, a downright fact may be told in a plain way; and we want downright facts at present more than anything else.—RUSKIN.

Original Communications.

THE ELEMENTS OF SURGERY.

BY JOHN CHIENE, M.D., F R C.S.E.

Surgeon to the Edinburgh Royal Infirmary, etc., etc.

LECTURE IV.—SUPPURATION.

The Varieties of Pus: Laudable, Specific—Whence the Pus Corpuscle —The Condition of Pus—Causes of Suppuration: Internal or Predisposing, External or Exciting—Aseptic and Septic : Mechanical and Chemical—Examples—Putrefaction—Definitive Cause—A Fermentation—The Germ Theory of Putrefaction.

The last lecture was devoted to the causes, symptoms, and treatment of inflammation.* Inflammation is the result of the application of an irritant, which may act directly on the tissues to which it is applied, or indirectly through the nervous system. I have already described the local phenomena. I have tried to show how the injured part recovers. The return to health takes place by resolution, absorption, and

---

* *Wildbad, Germany, August 30, 1879* —An apology on my part is certainly due to the Editor, for my delaying in forwarding to him this lecture in time for the September number. If any of his readers have noticed the break in the appearance of these lectures, it is but right that they should know who is to blame; perhaps they will forgive me when they note the place where this is written, and remember that it is a *health* resort of no mean repute.

organization. In connection with a wound, we term this the process of repair. If the action of the irritant is continued, if the inflammation is of a severe type, if the patient is weakly, or if the treatment is faulty, then recovery does not take place—a passage onwards and *downwards* takes place. The injured part can not stand still—either recovery or lócal death must take place. We have considered the recovery ; we have now to consider the local death. , There are three phases of this death — suppuration, ulceration, mortification — which differ only in degree.

Suppuration or the Formation of Pus.—It is not necessary to do more than mention the different varieties of pus. We have a standard pus,—viz., healthy or "laudable" (?) We have various departures from this standard ; hence the terms grumous, ichorous, serous, curdy, specific, etc. It will generally be found that the greater the tension the thicker is the pus. It is serous in those cases in which the tension is slight. Some surgeons look on pus as a something to be courted— on its appearance as by no means an evil ; hence the term "laudable." In a very limited sense they may, perhaps, be right. An epidemic of cholera or a war may be useful in clearing away the excess of the population ; an attack of diarrhœa may wash away some irritating material in the intestinal canal ; a bleeding from the nose may save one from an apoplexy ; a good sound thrashing may at times act beneficially, as also may a blister. These means are all *curative*, and require a preëxisting diseased condition ; and in so far, and no further, may we look on pus as laudable. Let us, however, never forget that the appearance of pus means a waste of the tissues—a local death—a corpuscular death-rate, excessive in amount, which is to be likened to the excessive death-rate still prevalent in certain districts. Whenever pus appears it means a diseased wound, as certainly as a high mortality in a town means something wrong with the drainage or water-supply. We now have sanitary inspectors who by *prophylaxis*, prevent a high mortality in districts under their charge. So also the surgeon can, by prophylactic measures

— for example, a drainage-tube, antiseptic precautions — prevent disease in wounds, and prevent the necessity for the appearance of pus. I again repeat that a wound will heal without pain or any uneasiness whatever, if it is properly treated, and that the process of recovery by which a wound unites is not a disease, and that even "laudable pus" itself should not be looked on as a necessary evil.

The term specific pus is often used—(it is a cloak to ignorance) ; pus with certain peculiarities, as, for example, in a gonorrhea caught in the usual way. How different this disease from a purulent urethritis—the result, it may be, of excessive sexual intercourse, or the rough passage of a bougie. The difference has yet to be traced ; it is a subject well worthy of study. One great difficulty in the investigation undoubtedly exists in the separation of the specific poison from the admixture of the causes of common putrefaction. Speaking from some clinical experience, I am inclined to the opinion that when a clap is cured in a few days, it must have been a simple urethritis, and not due to the planting of that specific something (undoubtedly endowed with life, and which may be called a germ), which gives rise to the common gonorrhea, which runs a course of four to six weeks as certainly as the crisis in typhoid fever is reached at the end of the third week.

It is not required that I should discuss here the question which must be solved by the pathologist,—Whence the pus corpuscle ? Is it the result of the proliferation of the cellular elements in the extra-vascular tissues of the inflamed part— a doctrine first enunciated by Goodsir, and elaborated by Virchow ? Is it, as Cohnheim and others teach, a white blood corpuscle, which has migrated from the vessels through the vessels walls ? Is it a result of the proliferation of the white blood corpuscles after their escape from the vessels ? That the white blood corpuscle does pass through the walls is an undoubted fact. That white blood corpuscles collect in inflamed tissues in great profusion is also very evident. What bearing these facts have on the origin of the pus corpuscle is still *sub judice*. In all probability both factors are at work.

In tissues, such as cartilage and cornea, the preëxisting cell elements, by their proliferation, supply the pus corpuscle; in vascular tissues, the migratory white blood corpuscle may be afterwards proved to be the pus corpuscle.

Pus, as we meet with it in the body, may be in one of three conditions—

(1.) *Aseptic,*—free from putrefaction, free from smell.

(2.) *Septic,*—putrid, with a more or less distinctly marked odour of putrefaction.

(3.) *Aseptic, but fetid,*—stinking.

The first condition is by far the most frequent—as in any unopened abscess. The second condition is met with in abscesses communicating with the external air, or with any of the canals in the body which contain the causes of the putrefaction. The third condition requires a word of explanation. In some abscesses (in the perineum, in the ischio-rectal fossa, and sometimes in the iliac fossa, in close proximity to the intestinal canal), when opened, the pus has an intensely disagreeable odour. If the pus from such an abscess is examined microscopically, bacteria may or may not be present. In 'some none are seen; and if, in such a case, the pus, with aseptic precautions, is inoculated into a flask containing a sterile solution of Darby's extract of meat (a most putrescible fluid), no result follows, showing conclusively that the fetid fluid does not contain living causes of putrefaction. The stinking pus may be, in all human probability is, the product of putrefaction, just as the smoke from an ignited charge of gunpowder is the product of the gunpowder. In those cases in which bacteria are found, on opening an abscess, there must be either a communication with a canal containing them, or it may be that they make their way from the blood-vessels in which, in certain diseased conditions, they are known to exist. The fact that the causes of putrefaction seem unable to pass through the tissues from the intestinal canal into an abscess cavity near it, while the products of putrefaction pass with ease, seems to be another proof that the causes of putrefaction are *particulate* and not in chemical solution.

Mr Lister proved this with regard to Edinburgh drinking water, and drove another nail into the coffin of the theory that the causes of putrefaction may be found by, nay arise out of, a combination of chemical atoms.

Causes of Suppuration.—Any cause which gives rise to the primary inflammation will, if persistent or powerful, give rise to suppuration. (See Causes of Inflammation, American Practitioner, August 1879, page 66.) An unsound constitution, an intense inflammation, the peculiar character of the inflammation, as in small-pox and some varieties of erysipelas, all rudimentary tissues as granulation tissue, may be taken as examples of *predisposing* causes to suppuration. They act from *within;* difficult to foresee, difficult to prevent, often interfering with the healing of a wound, they must never be forgotten. It is, however, different with the external causes acting from *without*—local in their action, directly applied to the wound, it is much more easy to defend the wound from such an attack.

It is to these external causes that I desire to direct special attention. A clear understanding of these causes of suppuration will go a good way towards clearing away many misunderstandings regarding "Lister's Antiseptic Method," and will illustrate the principles on which the practice is based; understand the principles—being understood, the practice will follow. There is nothing which Lister has laid more particular stress on than this—"Understand the principles on which I work before you practice the art." Much confusion has followed and many mistakes have been committed by those who have not studied these principles, and who have supposed that the purchase and use of a spray-producer, the enveloping of the limb in layers of antiseptic gauze, after dosing the poor wound with carbolic lotion from a large syringe, is all that is required in order to carry out Lister's system. How often do they find that the wound suppurates; it would be a wonder if it did not.

The conclusion is arrived at "the system is at fault." But the deduction is an erroneous one. Suppuration may have resulted from the abuse of the means used. Carbolic acid is an irritant; apply it in a strong solution, or with a daily persistence, as many do, and suppuration will result. *Carbolic acid is the evil of the antisceptic system.* The less that gets to our wounds, the better for our patients; it must be used, and we must understand why it is used before we can use it properly. The Lancet, in the early days of antiseptic surgery, when the system was in its infancy, said, "Lister's arguments are getting stronger and his solutions are getting weaker." Lister's arguments were getting stronger *because* his solutions were getting weaker. He was gradually finding out that the strength of the carbolic acid could be diminished, and that he still could prevent putrefaction. He was diluting the evil without interfering with its efficiency; he irritated the wound less,—the wound healed more kindly. Hence the good results were more constant. Suppuration, in consequence of the carbolic acid, occurred less frequently than in the early days when it was used in too concentrated a form. Take a wound in a healthy man; stitch it up closely, allow the effusions no room for escape—these effusions, at first serous, soon become purulent: here tension is the cause of the suppuration — the retained effusions act mechanically on the tissues, and by their pressure they cause increased irritation— suppuration is the result. Apply too tight a stitch, soon a circle of redness is seen round the stitch, and if the stitch is not cut a drop of pus forms, the tissues break down, the stitch is loosened; here the pressure of the too tight stitch acts mechanically and causes suppuration. Apply to an open wound any irritant, such as nitric, sulphuric or carbolic acid, and suppuration will occur; the irritant here is a chemical one. Lastly, neither stitch your wound nor apply any chemical irritant, and the wound may suppurate. Here the suppuration is due to some irritant which has also been applied to the wound. We may not see it, although its consequences are soon very evident. Putrefaction has occurred.

Lister first showed that putrefaction can be prevented.—He said prevent putrefaction, and you will prevent ONE of the

causes of suppuration. He never lost sight of the fact that
there are other causes of suppuration over which antiseptic
precautions have no direct control; and he also showed that
antiseptic precautions are a direct cause of one variety of
suppuration—that, namely, which is due to carbolic acid. He
adopted other means against these causes. He prevented
tension by the systematic use of a drainage-tube; he took
care take not to stitch his wound too tightly, knowing well the
evils of a tight stitch. He diluted the carbolic acid which
he used. He undoubtedly used it too strong at first, but
gradually he found out the strength which was necessary in
order to prevent putrefaction. At one time he used the
lotion too weak, but a few failures showed that he had gone
too far in his anxiety to avoid irritation. He felt his way,
and always keeping in view, on the one hand, the evil of too
free a use of the agent, and, on the other hand, never losing
sight of the great object he had in view in using it at all, he
at last struck the happy mean, whereby he was enabled to
prevent putrefaction, without at the same time applying too
powerful a local irritant. It is a misfortune, and no one
regrets it more than Lister himself, that the system ever
received the name of the "carbolic acid system."

Surgeons thought, and some still think, that the more
carbolic acid the better, forgetting that in this as in everything else—be it wine, be it food—we may have too much of
a good thing.

I have thought it right to make these remarks, because a
daily syringing of wounds is still called "Lister's treatment;"
say in an amputation at each dressing, or in an abscess after
opening, then suppuration occurs, and Lister has to bear the
blame. Not a drop of carbolic acid should reach the cavity
of a stump or an abscess cavity. The practice which will
soon receive our attention will illustrate this. I would just
ask why syringe your wound cavity? Is there anything mischievous in it that you want to destroy? Is it done to en-
courage healing? Is it used to wash away something hurtful?
It is for those who use it to answer these questions. Let it

be distinctly understood that Lister never advocated any such treatment. Carbolic acid must reach the wound during the operation. Carbolic acid lotion must be injected into the cavity of an accidental wound. After the drainage is introduced and the wound stitched up, then no more carbolic acid is to reach the cavity; it is hurtful to the healing process. If surgeons will inject carbolic acid at the daily dressings, then they must not say that they are following "Lister's method."

I have already mentioned the principal predisposing internal causes of suppuration. The external or exciting causes may be divided into the *asceptic* and the *septic.* Each of these divisions may again be divided into the mechanical and the chemical. We have, then, four varieties:

1. Asceptic mechanical. 2. Septic mechanical.
3. Asceptic chemical. 4. Septic chemical.

Examples of each will best illustrate the meaning of these terms. Sepsis is equivalent to putrefaction.

1. *Aseptic mechanical.*—A chemically clean bullet or a thorn; a non-putrid slough or blood clot; asceptic effusion in a wound, be it serum or pus; pus in an unopened abscess; a chemically clean silver stitch.

2. *Septic mechanical.*—A bullet or a thorn which carries in along with it the causes of putrefaction; a putrid slough or putrid blood-clot; putrid serum or pus in a wound; a silk stitch.

3. *Aseptic chemical.*—Iodine; sulphuric, nitric, carbolic acid; caustic potash, etc.

4. *Septic chemical.*—The products of putrefaction; sulphuretted hydrogen, ammonia and its compounds, carbonic acid, butyric acid, etc.

It is now necessary to say a few words regarding sepsis or putrefaction. Putrefaction is the series of changes which take place in substances containing nitrogen, when placed under the following conditions:

First, the vitality of the substance is gone or lowered; second, the substance is exposed to air or water at a temper-

ature between 212° Fahr. and 32° Fahr.   The vitality is gone
when we die.   The vitality is lowered when the tissues are
injured.   Exposure to the air is necessary.   Tinned meats do
not putrefy, because they are not exposed to the air.   Exposure to water is necessary.   Meat or fish which has been dried
is not prone to putrefaction.   If the substance is kept above,
or even near boiling point, it will not putrefy; so also if it is
frozen, it will not putrefy.

We have next to consider the cause of putrefaction.   If
we examine putrid matter, we find in it numbers of rod-like
bodies, which are termed bacteria.   These bacteria are
living; they have all the characteristics of life; they move, they
propagate like bodies; they require certain conditions for
their existence; they require more favourable conditions for
their growth and propagation.   They are seen in putrid
matter.   Are they the *cause* or the *result* of the change which
is termed putrefaction?   That they have something to do
with the process is very evident.   Putrefaction is allowed to
be a form of fermentation, and we may liken it to the well
understood variety of fermentation termed alcoholic fermentation.   If to sugar we add the yeast-plant—a living organism
—and keep it at a certain temperature, we find after a time
that the sugar is converted into alcohol and carbonic acid ;
these are the products of the fermentation.   We also find that
the yeast-plant is increased in quantity.   The yeast-plant
causes the change in the sugar, and it grows at the same time
that the change is taking place.   The bacterium is the yeast-
plant; the nitrogenous substance—say a solution of meat—
is the sugar.   The products of putrefaction—analogous to the
alcohol and carbonic acid—are sulphuretted, hydrogen, carbonic, butyric, valerianic acids, ammonia and its compounds,
etc.   These products are the septic chemical irritants, and
they result from the nitrogenous substance on the addition
of the bacterium, in the same way that the alcohol and
carbonic acid result from the sugar on the addition of the
yeast-plant.   The bacterium, then, is the cause and not the
result of the sepsis of putrefaction.

Another question is mixed up with the origin of putrefaction. Do these organisms arise from parents, or can they be formed *de novo?* It is now hardly necessary to consider this question, as almost all are now agreed that Pasteur was right when he asserted that these organisms arise from parents and that spontaneous generation is no more true of the bacterium than it is true of the yeast-plant, the maggot in meat, or the mite in cheese. When Lister proved that the causes of putrefaction are *particulate,* and not in chemical solution, he made a great advance, and his experiments, to my mind, go far to prove that life cannot arise without preëxisting life.

The *germ theory of putrefaction* asserts, first, that living organisms are the cause of putrefaction; second, that these organisms arise from parents; third, that they are planted in the substance which putrefies; fourth, that putrefaction is the result of the growth of these organisms in the substance which putrefies. Certain substances termed antisceptics interfere with this change, and they interfere by destroying the organisms which cause the change.

We have then to note that we have to deal with two factors, a living organism and a nidus for its life—a plant, and the soil in which it is planted. We will see, in the next lecture, that the growth of the organism can be interfered with in two ways. We may destroy the organism, or we may render the soil in which it grows unsuitable for its development. The object which the surgeon has in view may be likened to the daily work of the farmer in preventing weeds from growing on his land. The farmer either attacks and destroys the weeds or mixes something with his soil which will prevent the weeds from growing. This is the problem difficult for both surgeon and farmer, and the solution of this problem is antiseptic surgery. In the next lecture I will try and show that the surgery of the present day is a more or less perfect attempt to interfere with these organisms which cause putrefaction. The methods are very various. We will need to try and come to a decision as to which is the best.

THE AMERICAN PRACTITIONER.

DECEMBER, 1879.

Certainly it is excellent discipline for an author to feel that he must say all that he has to say in the fewest possible words, or his reader is sure to skip them; and in the plainest possible words, or his reader will certainly misunderstand them. Generally, also, a downright fact may be told in a plain way ; and we want downright facts at present more than anything else.—RUSKIN.

Original Communications.

THE ELEMENTS OF SURGERY.

BY JOHN CHIENE, M.D., F.R.C.S.E.

Surgeon to the Edinburgh Royal Infirmary, etc., etc.

LECTURE V.

CONTENTS. — Problems — The Prevention of Putrefaction—An Interference with the Growth and Development of the Causes of Putrefaction — Axioms — Germs Living — All Living Things require Certain Conditions for their Growth—These Conditions can be Interfered with—Vitality is their Poison : Non-Vitality is their Life—Examples—An Egg : An Apple— The Urethra and Bladder—Putrefactive Cystitis—The Danger of Tying in a Catheter— An Aseptic Catheter—The different Methods of Interference with the Cultivation of the Bacterium---Conclusion.

Having tried to show you reasons why we must accept the germ theory of putrefaction, we have now to attempt a solution of this problem,—How can we best interfere with the growth and development in our wounds of those organisms which are the cause of putrefaction? Keep prominently in view the two factors in the process—the organism which is planted and the tissue in which it is planted; put aside, for the moment, the method of treatment associated with Lister's name, and let us consider the different methods of treating wounds, and see if we can explain the good results met with on the broad principle laid down by Pasteur, which, if true in one case, must be true in all. The method adopted in the solution is antiseptic surgery in the widest sense of the term; the result of a solution of the problem is aseptic surgery. Note, in the first place, three axioms :—(*a*) Germs are living; (*b*) All living things, including germs, require for their growth and development certain conditions ; (*c*) These conditions can be interfered with.

(*a*) No one will deny that these particles (germs—bacterium) are living; they have a living principle; you see them growing, breeding, dying—all characteristics of life. Living things are divided into two great classes—animal and vegetable. The general impression is, that these organisms are vegetable. Being noxious, they may be considered as weeds ; and as a farmer does everything in his power to keep his land clear of weeds, so it is the surgeon's duty to prevent the weeds from appearing in his wounds.

(*b*) All living things require, for their growth and development, certain external conditions. I wish to separate growth and development from mere existence. A potato will exist on a dry stone floor ; it will not die, but no growth will take place, as it requires certain conditions which do not exist in the stone floor : but plant it in a suitable soil and supply it with moisture, and the result will be growth and development—a number of potatoes. There is, then, a difference between mere existence and growth, due to a change in the external conditions of the organism.

What are the conditions necessary for the growth and development of a bacterium ? Vitality in the soil in which they grow is their poison ; non-vitality is their life. Imagine a line A. B. Let it represent the average amount of vitality in the human race ; that is to say, the power which we have of resisting external influences. This power varies greatly in different individuals. One man can walk ten miles without fatigue ; another can not walk one. There is a mean vitality in the human race which may be represented by the line A. B. There are certain people whose vitality is above that line, and others whose vitality is below it. Above the line you have the area which acts as the poison to the bacterium ; below the line you have the area which is its life—an area in which it grows and develops. The further you go below the line the more suitable becomes the nidus, until at last you reach a point when all vitality has disappeared, when death takes place ; here the nidus is most suitable—here the bacterium flourishes. John Hunter's definition of life centres itself in this property of living things : — "the power of resisting putrefaction"—the power of preventing the growth of the bacterium. If I lay a bacterium on the back of my healthy hand, it may exist, it does not grow ; it has its analogue in the potato on the dry stone floor. If, however, I lay it in a wound or on an ulcerated surface, in which there is necessarily a lowering of the vitality, it there finds a suitable nidus— growth and development take place.

Let me illustrate the fact that vitality is their poison, by another example from John Hunter's work. In speaking of putrefaction, he illustrates the principle of life by the following curious and interesting fact : You are aware that the egg of the common fowl may be fecundated, or it may not ; that is to say, the former will, if placed in proper condition, grow and develop into a chicken—the latter will not. Now, if you place two such eggs—one fecundated, the other not—in an oven at a certain temperature, you will find, after a time, that a chicken comes forth from the fecundated egg, and that the albumen left in the egg is quite free from putrefaction. Break the other egg, and it will be stinking. You observe that the living principle in the fecundated egg prevented the growth and development of the mischief, while the absence of the living principle in the non-fecundated egg rendered the contents a fit nidus for the growth of the mischief.

Take another example : An apple, when it is bruised, will putrefy at the rotten part. The bruising of the apple destroys or lowers its vitality, and the result is, that the organism falling on that part finds a suitable soil for its growth—the bruised portion has lost the power of resisting external agencies.

The frugal housewife never stores fallen apples; they are at once converted into apple-tarts.

But let us take another example more closely connected with our bodies. I ask you now to think of a flask with a bend in the neck: you all possess it—the bladder and urethra. It may be compared to a glass flask with a long neck. There is this important difference, that in it the walls are living. In the bladder we have a putrescible fluid; this fluid is expelled three or four times in the twenty-four hours. At the opening of the urethra (at the mouth of the flask), the bacterium is present. Remember that the urethra is practically a capillary tube, the walls of which are separated by a layer of mucus; there is, however, plenty of room for the passage of these microscopic organisms. We have a most putrescible substance, at a comfortable temperature, for the growth of these organisms; and still as long as the walls of the urethra are healthy these organisms never pass up the urethra. This can be proved experimentally. Destroy with carbolic lotion the organisms at the opening of the urethra, and then make water into a glass flask which has been purified (with heat let us say), then close the mouth of the flask, and the urine will not putrefy. Make water, without washing the urethral opening with carbolic lotion, into a purified flask; close the mouth of the flask, and in forty-eight hours the urine will be loaded with bacteria. This proves that the bacterium does not exist in the healthy urethra. It also proves that the bacterium exists at the opening of the urethra. Take a capillary glass tube, which has been purified by heat, and after it cools introduce into it sterile albumen; close one end, and inoculate the other end with some fluid containing bacteria. In a few hours the albumen will become turbid, and this turbidity will gradually extend along the tube; the bacteria will be found throughout the whole length of the tube. They march up this tube; they don't do so up the healthy urethra. Why? Because the urethral walls are living—vitality is the poison of bacteria; the glass walls are dead—non-vitality is the life of bacteria. Let us pursue this subject a little further.

There is a common inflammation of the urethra—a gonorrhea. A not uncommon result of gonorrhea is cystitis. There are two kinds of cystitis. One due to an impression which passes along the sensory nerves to the spinal cord, and is reflected to the bladder, and an inflammation of the bladder is the result. Another variety of cystitis is due to the passage of the organisms along the urethra into the bladder; an inoculation of the urine takes place, and a form of cystitis, which may be termed purtrefaction, is the result: the urine is loaded with bacteria. The primary inflammation in the urethral walls lowers their vitality, and the mucus in the canal becomes at once a fit nidus for the growth of the organism. The urethral walls approach a condition analogous to the walls of the glass tube. Frequency of micturition in gonorrhea is a provision of. nature to wash away the mischief, and prevent its passage upwards into the bladder. Hence the value of diluents of barley-water, etc.; they act in the same way as a syringe in our wounds. And, in passing, I may say, that the urethral syringe should always be fitted with a fine piece of india-rubber tubing, which should be passed up the urethra beyond the tender inflamed area, so that the fluid injected may reach the urethra at the proximal side of the mischief, and escape along the sides of the india-rubber tube, washing the walls of the urethra in its passage in the same direction as the urine flows from the bladder.

If a catheter is tied into the bladder, then some antiseptic precaution must be taken to prevent the organisms from multiplying in the layer of mucus between the urethral wall and the outside of the catheter. If the orifice of the urethra is not washed with carbolic lotion, and a piece of gauze or boracic lint tied round the penis at the opening of the urethra, it will be found that after the catheter has been in position for two or three days the urine in the bladder is putrid, the organisms have passed along the sides of the catheter to the bladder in the same way as they passed along the glass tube. It will be well to change the catheter every second day, to wash it with carbolic lotion, oil it with carbolic oil (one to thirty), and reintroduce it. We must also remember that the catheter mechanically causes irritation; and after it has been in the urethra for a time, the walls of the canal inflame and the mucus becomes a fit nidus for the cultivation of the bacteria, still further increasing the chances of putrefactive cystitis. In the female, in consequence of the short urethra, the chances of cystitis from bacterial inoculation is more likely to occur than the male. This is, I believe, the explanation of the fact that the female is more liable to cystitis than the male; on the other hand, it is to be remembered that the bladder in the female is more thoroughly emptied, and that the mischievous particles are more easily washed out.

Why, then, if the dirty catheter is so mischievous, do we not have more frequently putrefactive cystitis? Because, in the great majority of cases, the bladder is healthy, and the next act of micturition empties the bladder completely, and the bacteria are washed away with the urine. If, on the other hand, we introduce a catheter into a bladder which is unhealthy, in which the walls are inflamed, in which the mucous membrane is corrugated and thickened, then some of the mischief may stick amongst the rugæ, or make its way into the softened mucous membrane, and may not be removed by the act of micturition, but remain attached to the walls, and lie in wait ready to inoculate the urine as it passes from the ureters into the bladder.

I have heard Mr Spence say, that it is not a good thing frequently to sound a man with stone before lithotomy. In other words, the more frequently this is done the greater is the chance of inoculating the urine in a bladder the seat of a stone, and which is almost certainly an unhealthy bladder with inflamed walls, in consequence of the mechanical irritation of the stone. Suppose a man suffers from prostatic enlargement, then we have to deal with a bladder which is never thoroughly emptied; and if we introduce with a dirty catheter mischief into such a bladder, then we have always a well of urine behind the prostate which contains putrid urine, and if there is any abrasion on the bladder, the putrid matter passes into the general circulation, and putrefactive product fever is the result. As long ago as 1869, Dr John Wyllie pointed out to me the clinical fact, that the first time you draw off the urine in a case of enlarged prostrate, the urine is clear, but in a few days the urine became thick and cloudy, with a copious sediment, and if this is examined with the microscope, bacteria are found.

The practical outcome of what I have said is, that an impure catheter should never be introduced into the bladder. In the healthy bladder it *may* do no harm; in the inflamed bladder, or in a case of enlarged prostrate, it will certainly be followed, sooner or later, depending on the frequency of introduction, by putridity of the urine. The catheter should be washed in one to twenty carbolic lotion (its surface and specially the cavity of the instrument), and oiled with one to thirty carbolic oil, before introduction. It will be well here to mention that "surgical kidney" is due to the passage of these organisms along the ureter into the pelvis of the kidney; the primary inflammation in the bladder causes inflammation of the ureter walls and walls of the pelvis of the kidney, and renders the urine in these situations a fit nidus for the development of the bacteria.

We have, then, the glass tube with the dead walls, the urethral walls with inflamed walls, and the urethral walls in a state of health, illustrating the proposition with which I began—that vitality is the poison, and non-vitality the life, of these organisms.

(*c*) I have now to direct your attention to the third axiom, that these conditions, which render easy the growth of bacteria, can be interfered with. I have now to speak of the various ways in which surgeons may interfere with the growth and development of these organisms in wounds, just as a farmer interferes with weeds on his farm. Good farming means no weeds, good surgery no organisms. The different methods of interference may be classed under the following heads :

(*A*). By removing them before they have had time to do harm.

(*B*). By removing the soil in which they flourish.

(*C*). By increasing the vitality of the tissues.

(*D*). By lowering the temperature of the tissues.

(*E*). By drying dressings.

(*F*). By preventing their deposit in a wound after an operation.

(*G*). By rapidly changing the wound from an absorbing to a casting off surface—preventing their entrance into the general circulation.

(*H*). By placing none in the wound at the time of the operation.

(*I*). By killing them.

(*A*). *By removing them before they have had time to do harm.* It is in this way that a syringe filled with simple water does good, simply washing away the mischief before it has had time to do harm. Just as a farmer roots up the young thistles before they have time to flower and spread by means of thistledown. Few surgeons now-a-days inject simple water; however much they cry out against "antiseptic surgery" (the word seems to act on some of the name as a red handkerchief does to a bull)—they use some antiseptic in their lotions. The fluid does its work as efficiently as the simple water as a flushing agent, and at the same time the mischief in the water is destroyed by the antiseptic.

(*B*). *By removing the soil in which they flourish.* By the removal of the serous and other discharges, which are a fit nidus for the cultivation of the mischief. The syringe acts in this way by rapid removal periodically performed. The drainage tube, the catgut or horse-hair drain, do the work gradually, and allow of no accumulation. Dr Thomas Keith at one time sucked up the discharges by means of a syringe introduced into a glass tube, which he introduced into the cavity of the pelvis at the time of the operation for ovariotomy. Some attain the same object by leaving the wound open, so that no discharge can accumulate. Other surgeons leave one part of the wound open. Others put on no dressing which is apt to retain the discharge.

(*C*). *By increasing the vitality of the tissues.* By bringing
your patient into as good a state of health as possible before
the operation; if the tissues are healthy, then the mischief
can not flourish—vitality is their poison. The local vitality
can be raised by preventing tension by means of the drainage
tube, or by any other of those means detailed under the
second head. Tension means inflammation; when a tissue
inflames its vitality is lowered, as I have already tried to show
in relation to the urethral walls when attacked by gonorrhea.
If the wound can be kept at rest, then inflammation is prevented. We owe this attack on unrest to the labours of the late
Mr Callender of St Bartholomew's Hospital. He carefully
supported his stumps on splints and pillows, preventing inflammation. It may be here well to note an attempt which is
now being made to render the body of the patient antiseptic
before operation, by the administration of carbolates. If this
can be managed *without poisoning our patients*, then we will
have a most efficient means of preventing putrefaction in our
wounds. As yet nothing definite can be said to have been
obtained.

(*D*). *By lowering the temperature of the tissues.* This can
be done by ice-bags or by an evaporating lotion. Keep the
part cool, and we prevent inflammation; we at the same time
render the tissues an unfit nidus for the growth of these organisms. In cold weather meat does not "go wrong."

(*E*). *By dry dressings.* At one time a very favourite way
was the application of dry lint. Moisture is one of the great
predisposing causes. The South Americans cut their meat into
thin strips, and hang it in the sun to dry, experience teaching
them that meat, when dry, will keep for an indefinite period.

It is often asked, How is it that union by the first intention
did occur in wounds before antiseptics were ever heard of, if
this mischief is so omnipresent? My answer is, that it was
by the various means alluded to under the above heads. No
special antiseptic was used ; the tissues were at a high pressure
rate of vitality—the mischief could do no harm. A germ in
the wound formed no nidus for its growth; it lay like a potato

on a stone floor, not dead but without those external conditions which it required for its growth and development. The
healthy tissues prevented its growth in the same way as the
walls of the healthy urethra prevent the mucus in the canal
from putrefying.

(*F*). *By preventing their deposit in a wound after an operation*, by intercepting them, by screening the wound, just as
you keep rabbits out of a garden by putting wire netting
along the fence. In Paris you will see this done by cotton-
wool. At the meeting of the British Medical Association in
1876, a glass box was shown in which the stump was placed
after an amputation.

(*G*). *By rapidly changing the wound from an absorbing to
a casting-off surface.* Mr Savory, in his eloquent address on
surgery, delivered before the British Medical Association this
year, advocates a well made poultice. How does this act?
By the heat engendered in the part the blood vessels dilate,
active tissue growth takes place, granulations speedily form,
and the walls of the wound cavity are changed from an absorbing surface to a casting-off surface. In other words, the
wounded surface is sealed, and the organisms which exist in
quantity on the wound surface find no entrance into the general circulation, and do no harm ; they are practically outside
the body. The poultice, from this point of view, acts as an
antiseptic agent, preventing one of the dangers of putrefaction, closing the wound in the most speedy manner.

(*H*). *By placing none in the wound at the time of the operation.* The hands of the operator and instrument are purified
with some antiseptic; the sponges are washed in an antiseptic
lotion.

(*I*). *By killing them.* This may be done in various ways :

First. You may kill them in the air of the room in which the
operation is performed, by means of disinfectants, chloride of
lime, Condy's fluid, carbolic or sulphurous acids. These
various substances, being volatile and poisonous, may be
placed in the room; and it is possible to understand that you
can prevent the living organisms from reaching the wound. I

can easily understand that in a hospital ward in which everything is soaked in carbolic lotion—basins of lotion, masses of
antiseptic gauze, and other carbolised materials lying about
—the atmosphere might be rendered a deadly mischief to the
poison.   In medicine disenfectants are used with the greatest
advantage, because, in a case of scarlet fever for instance, the
poison comes *from* the patient, and we destroy it when it
leaves the patient, and prevent those around him from harm.
Such means are useful in a putrid wound; in this case also
the poison comes *from* the patient, and we prevent the poison
from reaching other patients by surrounding the putrid wound
with some antiseptic.   It is, however, a different thing when
we make a wound.   With unbroken skin to deal with, our
business is to prevent the mischief from getting at the wound.
Disinfectants are not sufficiently local for the majority of
surgical patients.   I may illustrate the distinction between
the case of a patient suffering from typhus or scarlet fever,
and an operation, say, for the excision of a tumour, by the
following: You see fences round woods to prevent hares and
rabbits from getting out—to prevent them from reaching the
crops around the wood.   This is medicine.   You see fences
round gardens to prevent the rabbits from getting in.   This
is what we aim at in surgery.

    Second.   You may kill the organisms after they got in, by
the injection of some antiseptic lotion, as, for example, the
"hyperdistention" method of Callender acts in this way.
We must adopt this method when we are called to a wound
the result of an accident.   We can not help ourselves.   The
mischief has found an entrance, and it must be destroyed.
Therefore, in a case of compound fracture we inject carbolic
acid lotion into every nook and cranny, hoping to destroy the
mischief.   Of this we never can be sure ; some particle may
escape, and if left living, and it find a suitable nidus, it will
certainly grow and develop, and the wound will putrefy.
Moreover, we do this at a certain disadvantage, because all
antiseptics are irritants, and the tissues are, to a certain extent,
injured; and if we fail in destroying all the mischief, then

harm instead of good will follow. Therefore, when we inject a wound cavity, we must be thorough; half measures are worse than useless; the carbolic lotion acts as an irritant— the tissues are injured; and after the carbolic lotion which has been injected loses its antiseptic properties, the irritated tissues form a suitable nidus for the growth of the mischief which has escaped the poison.

Third. You may kill the mischief before it gets in. By performing the operation in an atmosphere which is poisonous. It is in this way the spray acts. We operate in a vapour of poison, so that any particle in the air in the immediate vicinity of the wound is destroyed, and when it reaches the wound it is dead, and can do no harm.

I have now directed your attention to the various ways in which you may interfere with the growth and development of these organisms, which I have tried to show are the cause of putrefaction. If I have been understood, it will be evident that we may combine these different methods with advantage, taking, as far as possible, the good out of each; avoiding, in so doing, the adoption of any method which is spoken of as " modified antiseptic," because we must be thorough, or we will be departing from the great principle which Lister has adopted as the basis of his work. We drain our wounds thoroughly in order to prevent tension. We keep them at rest, to prevent inflammation. We perform our operation under an efficient cloud of antiseptic spray, which is an active poison to the mischief. During the operation the surface of the wound is bathed by the antiseptic. No germs are intro- duced on the hands of the operator; he washes his hands in an antiseptic lotion; the instruments are purified in the same lotion. The drain is introduced, the wound stitched, and the dressing of some antiseptic material applied. The gauze dressing is so made that the antiseptic is constantly being given off at the temperature of the body. It is practically a continuance of the spray in the form of vapour, instead of a cloud of spray. When the discharges from the wound reach the edge of the dressing, then a new supply of the gauze must be applied. This is done under the spray. As the discharge becomes less and less, the dressing of the wound is less frequently required until the wound is healed. No carbolic acid should reach the cavity of the wound after the operation; it is not required; there is nothing mischievous in the wound; it would do harm, acting as an irritant.

I have now finished what I have to say on this subject. I have striven to show that all the best surgery of our day is more or less antiseptic; but besides this it is of the utmost importance that you should recognise the principles of the movement which is identified with Mr Lister's name, because if you do recognise these principles you will see the necessity of putting up with no half measures. I have heard much of " modified antiseptics." An English radical nobleman once asked his opponents who were pleading for moderate reform, what they thought of moderate chastity; and in this light I would view this question. There is no point at which we can stop on the line beginning with simple cleanliness and fresh air, and ending with all the contrivances which a trained scientific ingenuity can suggest for the prevention of putrefaction. Prevent putrefaction, and you prevent putrefactive irritation in wounds; but what is of infinitely more importance, you prevent the chance of the absorption of the products of putrefaction; you prevent that fever which is so deadly which I have called putrefactive product fever, and which has received various names — blood poisoning, septic poisoning, pyæmia, septicæmia. It matters little what it is called, if you but understand what it is and how to prevent it.

As a last word, I would not have you accept these doctrines without due inquiry and searching examination; but if you do accept them, you must be careful how you change your mind merely according to the results of your own practice. The foundations of the system are laid broad and deep on great principles—principles which I have tried to show have been for long followed empirically and in ignorance, until Lister grasped the idea and demonstrated the principle. No single individual has a right to throw it aside as worthless on the strength of the results of his own practice.

I have now, so to speak, prepared the way for the next lecture on the Practice of Antiseptic Surgery, which has been written by my late house surgeon and friend, Mr Francis Caird. I have asked him to write this lecture, as he has given much attention to the teaching of the subject to my students in the Practical Class, which forms an important part of the course of Systematic Lectures on Surgery, delivered by me to my class in the Edinburgh School of Medicine.

THE AMERICAN PRACTITIONER.

MARCH, 1880.

Certainly it is excellent discipline for an author to feel that he must say all that he has to say in the fewest possible words, or his reader is sure to skip them; and in the plainest possible words, or his reader will certainly misunderstand them. Generally, also, a downright fact may be told in a plain way; and we want downright facts at present more than anything else.—RUSKIN.

Original Communications.

THE ELEMENTS OF SURGERY.

BY JOHN CHIENE, M.D., F.R.C.S.E.

Surgeon to the Edinburgh Royal Infirmary, etc., etc.

LECTURE VI.—ON THE PRACTICE OF ANTISEPTIC SURGERY.

The various Antiseptics—Their Main Qualities—Carbolic Acid—Absolute Phenol—Watery, Alcoholic, and Oily Solutions—Antiseptic Gauze—Mackintosh Protective—Carbolized Catgut—Carbolized Silk—Steam Spray—Boracic Acid—Chloride of Zinc—Salicylic Acid—Drainage Tubes.

The practice of true antiseptic surgery (Listerism, as it has been termed)—that is, the keeping of a wound aseptic from first to last—requires not only a perfect understanding of the principles upon which the treatment is based, but also a careful consideration of the means employed to gain that end, and a thorough knowledge of the difficulties to be met with. This can only be gained by practice and experience; and gradually the slips and inaccuracies which may occur at first

disappear, and we become educated up to the necessary standard of excellence; so that what we tried to attain formerly
by unremitting attention and zeal we now obtain almost
instinctively and without effort. For a full account of antiseptic surgery we must seek the fountain-head in Lister's
writings, and much important matter may also be found in
the list of papers* here mentioned in addition to those already
referred to.

Our first duty will be to consider the various antiseptics at
present made use of. The main qualities required in an antiseptic are convenience, cheapness, and of necessity efficiency.
Carbolic acid so fulfils all those points that it still retains its
position at the head of the list, Mr Lister himself having given
up the use of thymol after a thorough trial. Carbolic acid
further is volatile—a property essential to any antiseptic in
use as a spray. The best form of acid to employ is the absolute phenol of Messrs Bowdler and Bickerdike Church, Lan-
cashire. Its advantages are that it has no objectionable

* LISTER : On a New method of Treating Compound Fractures, Abscess, etc.,
Lancet, vols. 1 and 2, 1867; On the Antiseptic Principle in the Practice of Surgery, Lancet, vol. 2, 1867 ; Illustrations of the Antiseptic System of Treatment
in Surgery, Lancet, November 1867 ; Antiseptic Treatment in Surgery, British
Medical Journal, vol. 2, 1868; On Ligature of Arteries on the Antiseptic System,
Lancet, vol. 1, 1869; On the Effects of the Antiseptic System of Treatment upon
the Salubrity of a Surgical Hospital, Lancet, vol. 1, 1870; On a Case of Com-
pound Dislocation of the Ankle, with Other Injuries, Lancet, vol. 1, 1870;
Address on Surgery, Plymouth, British Medical Journal, vol. 2, 1871 ; On a Case
Illustrating the Present Aspect of the Antiseptic System of Treatment in Surgery,
British Medical Journal, vol. 1, 1871 ; An Address on the Effects of the Antiseptic
Treatment upon the General Salubrity of Surgical Hospitals, British Medical
Journal, vol. 2, 1875 ; Demonstrations of Antiseptic Surgery before Members of
the British Medical Association, Ed. Medical Journal, vol. 1, 1875; On Recent
Improvements in the Details of Antiseptic Surgery, Lancet, vol. 1, 1875; Clinical
Lecture Illustrating Antiseptic Surgery, December 1879.

BISHOP : Article on Dressings, etc., in Swain's Surgical Emergencies.
CHAMPIONNIERE: Just. Lucas, Chirurgie Antiseptique.
KEITH : Results of Ovariotomy Before and After Antisceptics, British Medical
Journal, vol. 2, 1878.
LESSING: Article in the Deutsche Zeitschrift fur Chirurgie, book 3, page 402.
(Lister's Method in the Healing of Wounds.) A translation of this paper by Dr
Stirling, now professor in Aberdeen University, will be found in the Edinburgh
Medical Journal, March 1874.

G

odour, is readily soluble, and does not irritate the operator's
skin ; while the more crude and impure forms met with
are occasionally so disagreeable and harsh that some of the
German surgeons anoint their hands with vaseline before
beginning work, in order to obviate this inconvenience.

Among the various preparations of carbolic acid we may
first take up the solutions.   There are two watery solutions—
strong and weak.   The strong consists of one part of acid
crystals in twenty parts of water.   It is used for washing and
purifying the skin and instruments ; for soaking sponges,
drainage-tubes, and horse-hair ; and for the steam spray.   The
weak, which is half the strength of the strong—one part of
the crystals in forty parts of water—is required for washing
the sponges during an operation, for soaking the " deep dressing," and for dressing generally.   These lotions should be
filtered after making, and had better be kept in large, blue,
glass-stoppered jars carefully labelled.

An alcoholic solution of the strength of one part of the acid
in five of spirit of wine is employed for cleansing wounds seen
a few hours after injury, and specially for those cases in which
dirt and foreign matter have obtained access to the tissues.

There are two oily solutions.   The weak—one part of crystals in twenty of olive oil—is used for purifying and lubricating urethral bougies, sounds, and catheters immediately previous to their introduction ; the strong—of one part in ten—
for applying to exposed dead bone in situations where we can
not at once remove it, but have to leave it for some time *in
situ ;* for example, in necrosis of the flat bones of the skull-
cap.   In such cases a piece of lint soaked in the oil is laid on
the bare bone and covered with a piece of gutta-percha tissue.

Antiseptic gauze is prepared by charging unbleached
muslin of open texture with the following mixtue (New Formula, 1879) : crystallized carbolic acid, one part ; common
resin, four parts ; solid paraffin, four parts.   This last prevents
adhesiveness. Paraffin does not blend at all with carbolic acid
in the cold, and therefore simply dilutes the mixture of carbolic
acid and resin, without interfering in the least with the tena-

city with which the resin holds the acid. The acid is only given off in sufficient quantity when the gauze is moist and at the temperature of the human body.

To charge the gauze the paraffin and resin are first melted together in a water bath, after which the acid is added, and all are stirred together. We have now to diffuse this equably through the cotton cloth; and this requires, first, that the cotton be at a higher temperature than the melting-point of the mixture; and secondly, that it be subjected to pressure after receiving it. The gauze is therefore heated in a trough, and as layer after layer is turned over the hot mixture is squirted on by means of a large metal syringe furnished with a series of perforations at the end. Finally, a large heated block is allowed to descend, which accurately fills the trough and subjects its contents to pressure. The quantity of fluid mixture employed should be somewhat less in weight than the amount of gauze.

The prepared gauze is used for the large superficial dressing; in loose pieces for padding and dressing irregular surfaces; for bandages; and also when wet, wrung out of one-to-forty aqueous solution, for the " deep dressing."

Mackintosh consists of thin cotton cloth having a layer of india-rubber waterproofing on one side. This should be evenly applied and continuous, so that the material is quite impervious. There must be no pin-holes in it. We shall consider its use hereafter.

Protective is made of oiled silk, coated on both sides with a thin layer of copal varnish, which renders the silk impervious to the carbolic lotion. Over this again a fine layer of car-bolized dextrin is laid, which allows the one-to-forty lotion, into which the protective is dipped immediately before use, to wet, and so thoroughly purify the surface. The protective is neither aseptic nor yet antiseptic; hence the necessity of making it so before application. Its action is thus purely negative. It keeps the edges of the wound clean, moist, and free from the irritating action of the antiseptic employed, which, owing to the copal varnish, can not penetrate to the wound; allows discharge to escape readily from under it into the dressing; does not adhere, and so is easily removed when necessary.

Carbolized catgut is thus prepared: To twenty parts of carbolic acid crystals add two parts of water, and to this again add one hundred parts of olive oil. Place this mixture in a flask, and in this put several skeins of catgut. These should be kept, by means of a few glass marbles or rods, above the level of the watery deposit which occurs. Seal the flasks hermetically and set them aside in a cool place. The gut must not be used until five or six months after this, and the longer it has been prepared the better.

Carbolized silk is prepared by immersing a reel of silk in melted beeswax containing about one-tenth part of carbolic acid. The silk is drawn through a dry cloth as it leaves the hot fluid, to remove the superfluous wax.

All these various requisites should be kept by themselves apart from all other dressings—the gauze in a tin box; the silk in a stoppered glass jar; sponges, drainage-tubes, and horse-hair in wide-mouthed jars of one-to-twenty lotion; and the gut in its oil.

The various forms of steam sprays employed are constructed on the principle of Adam's steam inhaler. The boiler should be strong, dome-shaped, and furnished with a safety-valve. It is filled by an aperture situated at the lower level of the dome, so that in filling this space is left clear for steam alone. It is a disadvantage when the boiler is filled at the very summit; and in hospitals sufficient care is often not exercised, the steam dome is encroached upon, and a jet of boiling water is thus thrown out in place of spray. The steam-pipe, provided with a stop-cock and ball-joint, passes forward from the top of the dome and ends in a fine point, through which the steam rushes with great force. United to the under surface of the steam point, at an angle of forty-two degrees, we have the carbolic point continuous with the upper extremity of the india-rubber tube which leads up from the reservoir of one-to-twenty lotion. As the steam rushes out over the carbolic point it creates a vacuum, and the lotion thus sucked up is driven off in a fine cloud of vapour which covers an area large enough for any ordinary operation, and which is quite respirable, not wetting, and effective at a distance of at least four feet.

On arriving at a patient's house we fill the spray with boiling water up to the base of the dome; never above this; and so .we avoid the danger, as I have remarked, of having the upper point blockd by particles of dust carried along in the jet of water which would ensue were the boiler overfilled. We light the lamp, noting that the wick is in good order and that there is a sufficiency of spirit. We judge that steam is up if it escapes with great force, and if it has a distinctly blue colour, when we shut off all carbolic acid, which may readily be done by compressing the carbolic tube with the fingers, and so seeing steam alone. One has also the peculiar rushing sound, the smell and taste of the spray to guide him in ascertaining if all is in working order. A small filter formed of a piece of sponge, inserted into the lower extremity of the carbolic tube, and secured in position by means of a gauze cap, will prevent the lower point from getting choked with dirt, which, falling into the open jar of acid, may be sucked up, and so cause trouble. Should the spray cease working, we may unscrew the points and affix the reserve pair found in the hollow handle, the wound being meanwhile protected by a "guard" consisting of a rag or piece of gauze soaked in lotion. In this way operative procedure is not hindered, and the defaulting points may be seen to and cleaned out with a horsehair or fine silver wire at a more fitting time. This is an accident which hardly ever occurs in private practice.

The other antiseptics employed may now be discussed.

A solution of chloride of zinc (forty grains to the ounce of distilled water) was introduced by the late Campbell de Morgan. It is chiefly used to brush over the cut lips of incisions and wounds in regions which we can not hope to keep aseptic, as in excision of the upper-jaw or lateral lithotomy. We may leave our dressing of strips of lint soaked in this solution *in situ* for forty-eight hours, so potent in this salt; and in this way, thanks to its searching character and non-volatility, the pain and unrest of dressing is avoided, and a dangerous period, during which blood-poisoning from absorption might take place, is tided over. Considerable smarting and pain ensue after application, and this continues for a varying period, according to the temperament of the patient. The use of chloride of zinc for purifying ulcers will be referred to shortly.

Boracic or rather boric acid is used as lotion, lint, and ointment. It is non-volatile, very unirritating—in fact, the least so of all antiseptics—but is not at all searching. It may prevent, it can hardly eradicate putrefaction. The lotion of one part of the crystals in thirty parts of water is coloured red with litmus, and thus at a glance we may distinguish it from other lotions. It is used for moistening the boric lint and for washing sores.

The lint is prepared by soaking ordinary surgeon's lint in a boiling saturated solution of boric acid, coloured red with litmus. It is allowed to cool, the lint is hung up to dry, and the remaining fluid poured off and used as boric lotion. The lint is of a pink hue, and glitters with the soft, flat, micaceous crystals. In a similar manner we may charge bibulous paper or the paper lint introduced by Messrs Wyeth of Philadelphia. We moisten the boric lint with boric lotion before application, and this for the same reason as we also soak the deep dressing of gauze or the protective in carbolic lotion. The surface of the material may be covered with germs of all kinds; because the antiseptic is not acting. We destroy these organisms by our active lotion, and as the aseptic discharge finds its way afterward into the dressing it dissolves and sets free quite enough of the stored up agent to render itself also antiseptic.

Boric ointment may be prepared by rubbing up one part of finely-levigated boric acid in five parts of vaseline. It acts as a sort of antiseptic protective, and is specially useful in the treatment of wounds in the face, where it allows the discharge to escape, keeps the wound sweet, and never adheres.

An emulsion of salicylic acid in one-to-forty carbolic lotion was introduced by Mr Lister for the purpose of checking the chemical changes which may take place under dressings which have been left unchanged for sometime. These changes, due to a chemical action between the gauze and the discharges under it, the sweat, etc., give rise sometimes to a troublesome irritation and eruption, formerly dubbed *eczema carbolicum*. A very little salicylic cream smeared on the surface of the protective or deep dressing effectually disposes of this.

Having now gone sufficiently into detail as to the different materials used to secure antisepticism, our next lecture will be devoted to a consideration of the manner in which they are applied.

THE AMERICAN PRACTITIONER.

APRIL, 1880.

Certainly it is excellent discipline for an author to feel that he must say all that he has to say in the fewest possible words, or his reader is sure to skip them; and in the plainest possible words, or his reader will certainly misunderstand them. Generally, also, a downright fact may be told in a plain way; and we want downright facts at present more than anything else.—RUSKIN.

Original Communications.

THE ELEMENTS OF SURGERY.

BY JOHN CHIENE, M.D., F.R.C.S.E.

Surgeon to the Edinburgh Royal Infirmary, etc., etc.

LECTURE VII.—ON THE PRACTICE OF ANTISEPTIC SURGERY.

An Example—Duties of Instrument Clerk, Dresser, and Spray Clerk—Ligatures—Drainage Tubes—Catgut and Horsehair Drains—Closing Wound—Deep Sutures—Button Sutures—Application of Deep and Superficial Dressings—Change of Dressing—When to be done—Evidences of Failure—Septic Cases—Examples—Chloride of Zinc—Boric Lotion—Iodoform—Use of the Guard—Gutta Percha.

Let us now consider the application of these materials, and let us imagine the removal of a tumour from the region of the groin as one would see this operation performed in a public hospital with plenty of assistants. As this is one of the regions where the surgeon's care and ingenuity are specially taxed in keeping his wounds aseptic, it serves as a good example of the style of dressing required.

On the surgeon's right stands the table with instruments

and dressings. The gentleman who has charge of these arranges his instruments in a flat, shallow porcelain tray containing enough one-to-twenty carbolic lotion to cover them. A sheet of india-rubber may be laid in the bottom of the tray, so that the bistouries can not have their edges turned by coming in contact with the hard porcelain. The duties of the instrument clerk comprise attention to the surgical cleanliness of the forceps, saws, and other necessaries. These must be free from dirt, and their teeth thoroughly purified. Needles, sutures, and instruments must be carefully carbolised and passed to the surgeon through the cloud of spray. ·

On the surgeon's left is located the table with sponges, basins, and lotions. The dresser here removes as many sponges as he may require from the jar of one-to-twenty in which they are kept soaking, and places them for use in a basin half-filled with one-to-forty lotion. A second basin with a similar quantity of one-to-twenty is next provided, and into this he puts a couple of towels and one sponge.

The spray clerk takes his position where his spray can throw a suitable cloud over the field of operation without incommoding the surgeon, obstructing the view of the spectators, or obliging the patient to respire the antiseptic. His duty is to replenish the spray-bottle with one-to-twenty as required; to see that doors and windows are closed, so that no draught may undo all by blowing the spray cloud from off the wound. If ether is used as the anesthetic we must give it a wide berth, for fear of the flame of the lamp causing an explosion.

The patient being now quite anesthetised, the dresser hands the surgeon the basin with the one-to-twenty. The skin over and around the tumour is then well scrubbed with the sponge; the two towels, after having the superfluous lotion squeezed out of them, are laid the one over the genitals and the other so as to overlap the blanket which covers the upper part of the patient's body. . In this way a sort of antiseptic basis is provided, over which the spray plays, and on which we may lay sponges and instruments with safety during the operation.

If necessary the pubis may be shaved; and this, together with a preliminary purification, may be done before the patient leaves the ward.

The surgeon now washes his hands in the lotion—not a mere dip, as if he were afraid to carry the smell away with him, but a thorough cleansing, especially around the flexures and finger nails. The basin of one-to-twenty is then held during the operation close to the vicinity of the wound, so that the assistants may purify their hands or any instruments should they inadvertently be carried beyond the area of the spray.

The spray is now turned on, and the operation proceeds. The sponges are passed as required, wrung out of one-to-forty lotion; and as the dresser receives the dirty ones he squeezes them into a pail standing at hand, then washes them in the one-to-forty lotion, and wrings them dry as wanted. When the sponges are required very quickly a relay may lie on the towel covered by the spray; but on no account should we have a store of them lying exposed to the air before passing. One simply courts failure by so doing.

The ligatures may be cut as required. The surgeon is not at liberty to have a stock attached to his button-hole or to carry a dozen in his mouth, and yet claim to carry out most rigid antiseptic precautions.

The tumour has now been removed; the bleeding checked by ligatures, which are cut short. It remains to close the wound, prevent tension, and keep it aseptic.

Tension is abolished by the use of Chassaignac's drainage-tubes, catgut or horsehair drains. The tubes are introduced to the bottom of the wound. Their number and size can only be learned by experience; but the drainage can not be too free. Their outer ends must be flush with the level of the wound, and the two loops of silk turned back at right angles over the lips of the incision, and thus the tubes can not slip back into the cavity they drain. It is sometimes preferable to introduce the tubes after stitching up. The tubes act mainly by capillarity, in such cases the serum finding its way out between the clot and the walls of the tube; hence there is no advantage in keeping the tubes clear. The disadvantage of the tube is that we must dress the wound at times solely to shorten or remove the tubes, when otherwise things need not have been disturbed.

The catgut drain requires more care and discrimination in its use. It is serviceable in cases where we feel there will not be sufficient stimulus of any kind to cause suppuration; for while it drains away serum readily it can not convey pus. The catgut is slowly disintegrated and absorbed by the tissues, and thus requires no assistance in its removal ; hence those dressings are avoided which we must make in order to shorten or remove tubes when those are employed.

The drains should be formed of eight or twelve ply of gut tied in the middle ; and here we stitch it to the bottom of the wound with gut also, so that it may not be floated up to the top acting as a mere superficial drain, and leaving material to accumulate in the deeper parts and cause tension. The two free ends of the drain may be brought out at the extremities of the incision, or divided into three or four parts, which are brought out between the stitches.

Tubes and drains may be combined, as the surgeon sees fit; or he may use the tubes for the first two dressings, then insert drains instead, and look no more at the wound until such time as he hopes to find all healed, with the unabsorbed ends of the catgut lying on the top of the cicatrix. This one may readily carry out whenever the discharge is purely serous and small in quantity.

He now proceeds to close the wound. If it should be large and gaping one gets splendid results by the use of deep sutures, stitches of relaxation, or button sutures. A stout silver wire is carried through the integument about two inches or so beyond the incision, brought out at the wound, reintroduced and pulled out through the skin at a similar distance on the other side. The end of the wire has a flat leaden button or plate attached to it, and as the wire is now pulled tight and the superfluous wire with the needle attached cut off a second button is slipped on and the wire secured to it. These buttons hold the lips of the wound together in the same manner as one's fingers would act. Immediate union is now favoured by the introduction of numerous secondary stitches of coäptation., For these horsehair is preferable on account of its strength, elasticity, pliability, cheapness, and the ease with which it can be removed. It will be found advantageous to double the first twist of the reef-knot, and thus there will be no danger of the stitch relaxing while the final twist is being made. If a single hair is not strong enough, we may use two or three together.

It is hardly necessary to add, that all the sutures and needles must be duly purified, nor must the ends of the ligatures, etc., be allowed to touch any septic body as they are passed to the surgeon in the spray.

The whole operation has now been conducted under the spray antiseptically. We have now to maintain this aseptic condition. For this purpose the dressing is applied. It may be prepared beforehand. The surgeon lays a slip of protective, newly dipped in the one-to-forty carbolic lotion, over the lips of the wound, so that they are completely covered ; and with advantage we may also lay small pieces over the end of the drains, silk of tubes, and button stitches, these latter pieces being to prevent the gauze from adhering.

We now apply the deep-dressing ; and to understand its value we must bear in mind that dry carbolic gauze is not antiseptic. It gives off its acid at ordinary temperatures in such a small quantity and so slowly that is not even aseptic ; hence were we to apply dry gauze to recent wounds they would certainly in many cases putrify ; but we get over this difficulty by the use of a wet deep dressing consisting of three or four ply of gauze wrung out of one-to-forty lotion, laid over the protective, and extending for three or four inches beyond it all round. In this way we have destroyed any organisms which may have fallen on the surface of the gauze, the aseptic discharge is received by the active carbolic acid in the deep dressing, is there rendered antiseptic, and by the time it

reaches the large superficial dressing of dry gauze which has
now to be applied, the heat of the body has liberated so
much of the acid that there is now no danger of the discharge
putrefying.

After the deep dressing has been put on we may pad any
hollows with dry gauze, and perhaps put a special pad in that
region toward which the discharge will gravitate. Over this
is then laid the large superficial dressing, consisting of eight
ply of gauze, with a sheet of mackintosh interposed between
the seventh and eighth layers. One should note that the
glazed surface must always be turned inward, looking toward
the skin, and that it is slightly smaller than the square of the
gauze, so that it cannot protrude beyond it, for fear of so
infecting the discharge or screening it from view.

When the discharge soaks through the gauze it is at length
arrested by the mackintosh. It then makes its way toward
the margin of the dressing, taking with it in its course so much
carbolic acid that any organisms adherent to the polished
surface of the mackintosh are destroyed.

If there be a very copious flow of discharge it is possible
that all the carbolic acid may be washed out, and hence the
necessity for changing a dressing always for the first time
within twenty-four hours after the operation. The dressing is
secured by turns of a gauze bandage; these, from the manner
in which they cling and adhere, from their softness and
pliability, being very serviceable.

Whenever the four corners of the dressing are secured then
the spray may cease. Further security is gained by fasten-
ing the bandages to the four corners of the dressing with
safety-pins, and then the bandages to one another only,
where they cross. On no account allow a pin to perforate
the mackintosh.

For restless patients, and in cases which require dressing
but once a-week or less, a broad elastic bandage over all
gives both patient and surgeon much comfort, as any chance
of the dressing slipping is thus almost completely avoided.

The patient now lies undisturbed for the next four and

twenty hours, when we change the dressing for the first time. We now require a deep and superficial dressing as before, and the dresser has also in readiness protective, a basin containing one-to-forty lotion, deep dressing, and two rags, one of fine texture known as a guard, and a second coarser to swab with.

The spray is seen to be in order and in position. The surgeon now removes the pins and elastic, cuts the guaze bandage, and washes his hands in the lotion; the patient, if necessary, assisting meanwhile by pressing a hand on the dressing to keep it steadily in position. The spray is turned on, and the surgeon proceeds to lift up that corner of the dressing which is nearest the spray, so that the cloud may be directed into the angle between the dressing and the skin. The deep dressing and protective are now similarly removed, and we may gently wipe up any serum which may obscure our view of the wound. Should we desire to ascertain if there is retained discharge we cover the wound with the wet guard and then make use of gentle pressure with the fingers. The guard effectually protects the wound should the spray cease working or should any current of air turn the cloud aside; and again, it is a matter of the greatest moment to have only purified air in the vicinity when we relax pressure, as a regurgitation into the wound must ensue. The new protective is now dipped and applied; the deep and superfical dressings follow as before.

And now the question arises, When are we to dress again? This is settled either by length of time (a dressing should not be left on for more than about twelve days), or by appearance of discharge at the edge of the dressing. The discharge is best seen as a stain on the clean draw-sheet on which the patient lies, and the nurse has strict injunctions never to change the draw-sheet until it has been examined by the surgeon at daily visit.

It will be found that we no longer require to dress frequently; that intervals of three, four, and more days may elapse, until finally we find all healed. The event of any abnormal circumstance, as a rise of temperature, pain, or discomfort, even although minus discharge, will oblige us to dress without delay.

At these future dressings it will be necessary to attend to the drainage-tubes and stitches. The tubes should be shortened to the extent of about a quarter or half an inch, according as they are seen to be pushed out by the contraction of the tissues around them; and frequently they must not only be shortened, but we must substitute others of smaller caliber in their place, or use catgut drains—points which can only be ascertained by experience. The removal of stitches should present no difficulty. As regards button sutures, the surgeon seizes one button with his forceps, and gently pulls it upward, at the same time passing a strong pair of scissors beneath, so that as soon as the slightly curved portion of the wire has been withdrawn from the skin he may divide it. The other button and wire may now be withdrawn easily and painlessly.

When we open an abscess we must proceed as before to purify the skin and surroundings. The incision made should only be large enough to admit the drainage-tube or the finger, previously carefully cleansed, if it be deemed necessary to explore the cavity. Protective is not necessary.

In the case of psoas and lumbar abscesses, where the patients are long in bed, we must shorten the tubes slowly and carefully, never losing patience. A puckering in one of the lips of the wound points to a cicatrisation of the tissues, and indicates that all is going well.

It may now be noticed, that never once have we flushed our wounds with carbolic acid; never once have we employed a syringe. Our whole aim has been to admit as little as possible of the irritating antiseptic to our wounds. We use it merely to act as external agents, our wounds being left, as it were, subcutaneous, the protective keeping out the carbolic acid, and the dressing acting like the healthy skin in preventing the entrance of putrefactive organisms.

What are the symptoms and signs that enable us to say

we have failed in our object? The symptoms are many; individually no one is absolutely certain; collectively they are so. Of signs we have but one, and this one, plus a few of the symptoms, enables us to say that our wounds are or are not septic.

The chief symptoms shown by the patient are those of general or local constitutional disturbance, such as rise of temperature, inflammation of the wound, and suppuration; and as regards the dressing, we at once note that it stinks, and that the protective has become blackened by the sulphuretted hydrogen of putrefaction acting on the litharge with which the protective is prepared.

The occurrence of *all* the above would at once lead us to state that our purpose had failed; the occurrence of any *one* of them would only rouse our suspicion that all was not well, and would lead us to search for the special cause at work.

Thus the rise of temperature might yield to a dose of castor oil; the local redness and pain might subside on dividing a tight stitch which was giving rise to tension; the suppuration might cease on removing some foreign body causing irritation by its presence, such as a scale of dead bone; the stink of the dressing would come to an end when we applied a little salicylic acid; and the blackening of the protective might be due to the use of india-rubber drainage-tubes prepared with sulphur.

Any of the above symptoms combined with the one certain sign—the presence of *living, moving bacteria*—at once enable us to state that we have failed, when the sooner we resort to open treatment with antiseptics the better. We may ascertain the presence of bacteria best by examining a little discharge removed at the time of dressing from the *under-*surface of the protective; for here we find the organisms, if any are present, more lively, and so better marked than under the gauze. A power of three hundred and fifty is enough; and we need not be disturbed by the appearance of clustered oil-granules, or indulge in a vain hunt for micrococci, since it has been abundantly demonstrated that the latter have nothing whatsoever to do with putrefaction. We look only for well-marked bacteroidal rods. See paper by W. Watson Cheyne, Lancet, May 17, 1879.

Septic cases we may divide into two classes—recent, embracing fractures and wounds which, having been exposed to the air, are liable to become putrid before long; and secondly, chiefly joint-diseases, old standing, with broken skin and sinuses.

As regards the former, let us imagine a fracture of the leg due to a crush. The bone protrudes and is dirty; the accident occurred a few hours ago; hemorrhage has been controlled by means of a tourniquet. We expose the wound under the spray, and after securing the vessels with carbolized gut at once attempt to convert the compound into a simple fracture. The skin is purified with one-to-twenty lotion, which is also injected into the recesses of the wound by means of a gum-elastic catheter attached to a syringe. In this way the wound is thoroughly washed out; but care must be taken not to use the syringe forcibly, otherwise we may injure healthy tissue, the acid may be sent up the sheaths of tendons, and so sloughing will ensue. The protruding piece of bone is scrubbed with a nail-brush dipped in one-to-five alcoholic solution, so that the dirt which has been rubbed into it may be rendered completely inert. The bones are now laid in position, drainage-tubes inserted, and the dressing put on, the limb being left with the most suitable form of retentive apparatus applied externally. Within the next forty-eight hours we shall learn, from the behaviour of the wound, whether our efforts have been successful or not. Future dressings follow the ordinary rules.

In scalp-wounds we purify the wound and surrounding skin with one-to-twenty carbolic acid, cutting the hair close, insert a catgut drain, and stitch up. Such cases are remarkable for their rapidity of healing, the troubles of erysipelas and inflammation being quite unknown among a class of out-patients who are not the most attentive.

Let us now deal with an old-standing putrid-joint case complicated with sinuses. Our first duty is to remove all the putrid tissues we can, and this we endeavour to attain by scraping out the loose lining granulations from the sinuses with one of Volkmann's sharp spoons. In this way we hope to get rid of the unhealthy septic material, and leave behind only sound aseptic tissue. The sinuses are now injected with the chloride of zinc solution, and the operation continued as if with unbroken skin, a final purification with the zinc salt being made use of before the dressing is applied.

In dealing with ulcers we first purify the skin with one-to-twenty lotion, and then the ulcer itself is rubbed with a piece of lint saturated with the chloride of zinc. A piece of protective of the same shape as the sore, and about an eighth of an inch larger all round, is then dipped in boric lotion to purify it, and applied; while over this again we lay two layers of boric lint wrung out of the lotion. The lint must overlap the protective for an inch all round at the very least, and in cases where a copious discharge is anticipated a special pad of the lint may be applied at the most dependent part to receive it. A bandage from below upward completes the dressing, and the patient keeps the limb at rest until such time as discharge shows itself. Then a new dressing is applied. We wash the sore and adjoining skin with boric lotion, dip the protective and lint in clean lotion, and proceed as before.

The chloride of zinc causes a good deal of smarting and uneasiness, but the subsequent abolition of pain and smell is so marked that should the ulcer again become putrid patients request a second application of the solution.

We may also purify our ulcers by sprinkling them with powdered iodoform after washing with one-to-twenty carbolic lotion, and dress as above. The pain caused by the zinc may be avoided and the ulcer purified by using several poultices of boric lint; that is to say, apply the layers of moist lint and over them a large sheet of gutta-percha, and then bandage.

Continue this treatment for some days, and then dress with protective and lint.

Boric acid being a very mild, non-volatile, and non-penetrating antiseptic, is only suited for superficial wounds. It is, however, the least irritating antiseptic we possess.

The application of antiseptics in cases of fistula, etc., will be stated in the course of lectures.

In private practice one finds the carrying out of antiseptic detail even less troublesome than in hospital. The spray is not so liable to get out of order, since it never changes hands. We do not make use of so many assistants ; nor yet have we the convenience and benefit of bystanders to consider.

The surgeon may carry in his spray-bag a small supply of crystals of carbolic acid, so that he has practically a great quantity of lotion in a very small space. He has also sponges; but the dressings and lotion are usually found in readiness at the patient's house. While the patient is being anesthetised one gets the spray in order, arranges instruments and dressings. The spray during the operation stands on a small table in a convenient position, and requires but little attention. Should the carbolic lotion in the spray-bottle become exhausted, or should it be necessary to shift the position of the spray, then the surgeon merely lifts his guard out of the lotion, covers the wound with it, and so puts things to rights. Instruments may lie on a large plate, or in a tumbler of water, their points being saved contact by a cake of india-rubber laid over the bottom. Sponges of course are taken in hand by an assistant or nurse.

The future dressings are managed as in hospital. A daily visit is not required, since by means of a post-card the patient may send word to the surgeon should any discharge appear or discomfort be experienced.

www.ingramcontent.com/pod-product-compliance
Lightning Source LLC
Chambersburg PA
CBHW022047210326
41519CB00055B/1104